Food Safety and Quality

Food Safety and Quality

Edited by
Edmund Parker

Larsen & Keller
www.larsen-keller.com

Food Safety and Quality
Edited by Edmund Parker
ISBN: 978-1-63549-125-8 (Hardback)

Larsen & Keller
Published by Larsen and Keller Education,
5 Penn Plaza,
19th Floor,
New York, NY 10001, USA

Cataloging-in-Publication Data

Food safety and quality / edited by Edmund Parker.
p. cm.
Includes bibliographical references and index.
ISBN 978-1-63549-125-8
1. Food--Safety measures. 2. Food adulteration and inspection. 3. Food contamination.
4. Food--Quality. 5. Food industry and trade--Appropriate Technology.
I. Parker, Edmund.

TX531 .F66 2017
664.07--dc23

Printed and bound in the United States of America.

For more information regarding Larsen and Keller Education and its products, please visit the publisher's website www.larsen-keller.com

Table of Contents

Preface

As the food industry is experiencing a boom, new players are entering the market every day. This makes the process of checking the quality of food difficult. Therefore, new laws and regulations are being passed by the government to ensure the safety and quality of food. To provide the best preparation, handling and storage is referred to as food safety. Food safety and quality is a field of science that has undergone rapid development over the past few decades. This book is designed to provide students with knowledgeable insights about the field of food safety and food quality. Most of the topics introduced in this text cover new techniques and the applications of this subject. Coherent flow of topics, student-friendly language and extensive use of examples make this book an invaluable source of knowledge.

Given below is the chapter wise description of the book:

Chapter 1- The proper handling, preparation and storage of food in ways that prevent illnesses caused by food items is food safety. Food quality on the other hand is the quality of food, which is accepted by the consumer. This chapter is an overview of the subject matter incorporating all the major aspects of food safety and food quality.

Chapter 2- The measures taken for food safety are the preservation of food, storage of food and food defense. While food defense is the protection of food products from adulteration by biological, chemical and physical agents, food preservation involves the prevention of bacteria, fungi and also slows the oxidation of fats. This chapter helps develop a comprehensive understanding on food safety.

Chapter 3- Some of the methods of preservation related to food safety are food drying, refrigeration, frozen food, pickling and antioxidants. Food drying is a method in which food is dried, which stops the growth of bacteria and yeasts. Preservation of food by freezing is also a common practice, freezing food preserves it from the time it is prepared, till the time it is eaten. This chapter discusses the methods of food safety in a critical manner providing key analysis to the subject matter.

Chapter 4- Food safety is the basic concern of humans and with evolving times, better techniques have been introduced in this regard. Some of the modern techniques are pasteurization, vacuum packing, food irradiation, bio preservation and acidity regulator. The chapter serves as a source to understand these major categories.

Chapter 5- Contaminated food has certain effect on people; the effects discussed in this chapter are foodborne illness, food allergy, diarrhea, nausea and vomiting. Foodborne illness is any illness caused from contaminated food, by bacteria, viruses, or parasites that contaminate food and the symptoms often include vomiting, fever, and may also include diarrhea. The abnormal immune response to food is called food allergy, its symptoms usually include swelling of the tongue, vomiting, trouble breathing or low blood pressure. This chapter helps the reader in understanding the illnesses and problems that can be caused by food contamination.

Chapter 6- The inspection of food to make sure its fresh and to confirm its market value is food grading whereas the process of adding micronutrients to food items is food fortification. This chapter elucidates on topics such as, food grading, food fortification, optical sorting and food intolerance.

At the end, I would like to thank all those who dedicated their time and efforts for the successful completion of this book. I also wish to convey my gratitude towards my friends and family who supported me at every step.

Editor

1

Introduction to Food Safety and Food Quality

The proper handling, preparation and storage of food in ways that prevent illnesses caused by food items is food safety. Food quality on the other hand is the quality of food, which is accepted by the consumer. This chapter is an overview of the subject matter incorporating all the major aspects of food safety and food quality.

Food Safety

Food safety is a scientific discipline describing handling, preparation, and storage of food in ways that prevent foodborne illness. This includes a number of routines that should be followed to avoid potentially severe health hazards. In this way food safety often overlaps with food defense to prevent harm to consumers. The tracks within this line of thought are safety between industry and the market and then between the market and the consumer. In considering industry to market practices, food safety considerations include the origins of food including the practices relating to food labeling, food hygiene, food additives and pesticide residues, as well as policies on biotechnology and food and guidelines for the management of governmental import and export inspection and certification systems for foods. In considering market to consumer practices, the usual thought is that food ought to be safe in the market and the concern is safe delivery and preparation of the food for the consumer.

Food can transmit disease from person to person as well as serve as a growth medium for bacteria that can cause food poisoning. In developed countries there are intricate standards for food preparation, whereas in lesser developed countries the main issue is simply the availability of adequate safe water, which is usually a critical item. In theory, food poisoning is 100% preventable. The five key principles of food hygiene, according to WHO, are:

1. Prevent contaminating food with pathogens spreading from people, pets, and pests.
2. Separate raw and cooked foods to prevent contaminating the cooked foods.
3. Cook foods for the appropriate length of time and at the appropriate temperature to kill pathogens.
4. Store food at the proper temperature.
5. Do use safe water and safe raw materials.

ISO 22000

ISO 22000 is a standard developed by the International Organization for Standardization dealing with food safety. This is a general derivative of ISO 9000. ISO 22000 standard: The ISO 22000 international standard specifies the requirements for a food safety management system that involves interactive communication, system management, prerequisite programs, HACCP principles.

Incidence

A 2003 World Health Organization (WHO) report concluded that about 30% of reported food poisoning outbreaks in the WHO European Region occur in private homes. According to the WHO and CDC, in the USA alone, annually, there are 76 million cases of foodborne illness leading to 325,000 hospitalizations and 5,000 deaths.

Regulatory Agencies

WHO and FAO

In 2003, the WHO and FAO published the Codex Alimentarius which serves as an guideline to food safety.

However, according to Unit 04 - Communication of Health & Consumers Directorate-General of the European Commission (SANCO): "The Codex, while being recommendations for voluntary application by members, Codex standards serve in many cases as a basis for national legislation. The reference made to Codex food safety standards in the World Trade Organizations' Agreement on Sanitary and Phytosanitary measures (SPS Agreement) means that Codex has far reaching implications for resolving trade disputes. WTO members that wish to apply stricter food safety measures than those set by Codex may be required to justify these measures scientifically." So, an agreement made in 2003, signed by all member states, inclusive all EU, in the codex Stan Codex 240 – 2003 for coconut milk, sulphite containing additives like E223 and E 224 are allowed till 30 mg/kg, does NOT mean, they are allowed into the EU. Only for the latter the EU amended its regulations with (EU) No 583/2012 per 2 July 2012 to allow this additive, already used for decades and absolutely necessary.

Australia

Food Standards Australia New Zealand requires all food businesses to implement food safety systems. These systems are designed to ensure food is safe to consume and halt the increasing incidence of food poisoning, and they include basic food safety training for at least one person in each business. Food safety training is delivered in various forms by, among other organisations, Registered Training Organizations (RTOs), after which staff are issued a nationally recognised unit of competency code on their certificate. Basic food safety training includes:

- Understanding the hazards associated with the main types of food and the conditions to prevent the growth of bacteria which can cause food poisoning and to prevent illness.
- Potential problems associated with product packaging such as leaks in vacuum packs, damage to packaging or pest infestation, as well as problems and diseases spread by pests.

- Safe food handling. This includes safe procedures for each process such as receiving, re-packing, food storage, preparation and cooking, cooling and re-heating, displaying products, handling products when serving customers, packaging, cleaning and sanitizing, pest control, transport and delivery. Also covers potential causes of cross contamination.
- Catering for customers who are particularly at risk of food-borne illness, as well as those with allergies or intolerance.
- Correct cleaning and sanitizing procedures, cleaning products and their correct use, and the storage of cleaning items such as brushes, mops and cloths.
- Personal hygiene, hand washing, illness, and protective clothing.

Food safety standards and requirements are set out at the national level in the Food Standards Code, and brought Into force in each state by state-based Acts and Regulations. Legislation means that people responsible for selling or serving unsafe food may be liable for heavy fines.

China

Food safety is a growing concern in Chinese agriculture. The Chinese government oversees agricultural production as well as the manufacture of food packaging, containers, chemical additives, drug production, and business regulation. In recent years, the Chinese government attempted to consolidate food regulation with the creation of the State Food and Drug Administration in 2003, and officials have also been under increasing public and international pressure to solve food safety problems. However, it appears that regulations are not well known by the trade. Labels used for "green" food, "organic" food and "pollution-free" food are not well recognized by traders and many are unclear about their meaning. A survey by the World Bank found that supermarket managers had difficulty in obtaining produce that met safety requirements and found that a high percentage of produce did not comply with established standards.

Traditional marketing systems, whether in China or the rest of Asia, presently provide little motivation or incentive for individual farmers to make improvements to either quality or safety as their produce tends to get grouped together with standard products as it progresses through the marketing channel. Direct linkages between farmer groups and traders or ultimate buyers, such as supermarkets, can help avoid this problem. Governments need to improve the condition of many markets through upgrading management and reinvesting market fees in physical infrastructure. Wholesale markets need to investigate the feasibility of developing separate sections to handle fruits and vegetables that meet defined safety and quality standards.

European Union

The parliament of the European Union (EU) makes legislation in the form of directives and regulations, many of which are mandatory for member states and which therefore must be incorporated into individual countries' national legislation. As a very large organisation that exists to remove barriers to trade between member states, and into which individual member states have only a proportional influence, the outcome is often seen as an excessively bureaucratic 'one size fits all' approach. However, in relation to food safety the tendency to err on the side of maximum protection for the consumer may be seen as a positive benefit. The EU parliament is informed on food safety matters by the European Food Safety Authority.

Individual member states may also have other legislation and controls in respect of food safety, provided that they do not prevent trade with other states, and can differ considerably in their internal structures and approaches to the regulatory control of food safety.

From 13 December 2014, new legislation - the EU Food Information for Consumers Regulation 1169/2011 - require food businesses to provide allergy information on food sold unpackaged, in for example catering outlets, deli counters, bakeries and sandwich bars.

France

Agence nationale de sécurité sanitaire de l'alimentation, de l'environnement et du travail (anses) is a French governmental agency dealing with food safety.

Germany

The Federal Ministry of Food, Agriculture and Consumer Protection (BMEL) is a Federal Ministry of the Federal Republic of Germany. History: Founded as Federal Ministry of Food, Agriculture and Foresting in 1949, this name did not change until 2001. Then the name changed to Federal Ministry of Consumer Protection, Food and Agriculture. At the 22nd of November 2005, the name got changed again to its current state: Federal Ministry of Food, Agriculture and Consumer Protection. The reason for this last change was that all the resorts should get equal ranking which was achieved by sorting the resorts alphabetically. Vision: A balanced and healthy diet with safe food, distinct consumer rights and consumer information for various areas of life, and a strong and sustainable agriculture as well as perspectives for our rural areas are important goals of the Federal Ministry of Food, Agriculture and Consumer Protection (BMELV). The Federal Office of Consumer Protection and Food Safety is under the control of the Federal Ministry of Food, Agriculture and Consumer Protection. It exercises several duties, with which it contributes to safer food and thereby intensifies health-based consumer protection in Germany. Food can be manufactured and sold within Germany without a special permission, as long as it does not cause any damage on consumers' health and meets the general standards set by the legislation. However, manufacturers, carriers, importers and retailers are responsible for the food they pass into circulation. They are obliged to ensure and document the safety and quality of their food with the use of in-house control mechanisms.

Hong Kong

In Hong Kong SAR, the Centre for Food Safety is in charge of ensuring food sold is safe and fit for consumption.

India

Food Safety and Standards Authority of India, established under the Food Safety and Standards Act, 2006, is the regulating body related to food safety and laying down of standards of food in India.

New Zealand

The New Zealand Food Safety Authority (NZFSA), or Te Pou Oranga Kai O Aotearoa is the New Zealand government body responsible for food safety. NZFSA is also the controlling authority for imports and exports of food and food-related products. The NZFSA as of 2012 is now a division of the Ministry for Primary Industries (MPI) and is no longer its own organization.

Pakistan

Pakistan does not have an integrated legal framework but has a set of laws, which deals with various aspects of food safety. These laws, despite the fact that they were enacted long time ago, have tremendous capacity to achieve at least minimum level of food safety. However, like many other laws, these laws remain very poorly enforced. There are four laws that specifically deal with food safety. Three of these laws directly focus issues related to food safety, while the fourth, the Pakistan Standards and Quality Control Authority Act, is indirectly relevant to food safety.

The Pure Food Ordinance 1960 consolidates and amends the law in relation to the preparation and the sale of foods. All provinces and some northern areas have adopted this law with certain amendments. Its aim is to ensure purity of food being supplied to people in the market and, therefore, provides for preventing adulteration. The Pure Food Ordinance 1960 does not apply to cantonment areas. There is a separate law for cantonments called "The Cantonment Pure Food Act, 1966". There is no substantial difference between the Pure Food Ordinance 1960 and The Cantonment Pure Food Act. Even the rules of operation are very much similar.

Pakistan Hotels and Restaurant Act, 1976 applies to all hotels and restaurants in Pakistan and seeks to control and regulate the rates and standard of service(s) by hotels and restaurants. In addition to other provisions, under section 22(2), the sale of food or beverages that are contaminated, not prepared hygienically or served in utensils that are not hygienic or clean is an offense. There are no express provisions for consumer complaints in the Pakistan Restaurants Act, 1976, Pakistan Penal Code, 1860 and Pakistan Standards and Quality Control Authority Act, 1996. The laws do not prevent citizens from lodging complaints with the concerned government officials; however, the consideration and handling of complaints is a matter of discretion of the officials.

South Korea

Korea Food & Drug Administration

Korea Food & Drug Administration (KFDA) is working for food safety since 1945. It is part of the Government of South Korea.

IOAS-Organic Certification Bodies Registered in KFDA: "Organic" or related claims can be labelled on food products when organic certificates are considered as valid by KFDA. KFDA admits organic certificates which can be issued by 1) IFOAM (International Federation of Organic Agriculture Movement) accredited certification bodies 2) Government accredited certification bodies – 328 bodies in 29 countries have been registered in KFDA.

Food Import Report: According to Food Import Report, it is supposed to report or register what you import. Competent authority is as follows:

Product	Authority
Imported Agricultural Products, Processed Foods, Food Additives, Utensils, Containers & Packages or Health Functional Foods	KFDA (Korea Food and Drug Administration)
Imported Livestock, Livestock products (including Dairy products)	NVRQS (National Veterinary Research and Quarantine Service)
Packaged meat, milk & dairy products (butter, cheese), hamburger patties, meat ball and other processed products which are stipulated by Livestock Sanitation Management Act	NVRQS (National Veterinary Research and Quarantine Service)
Imported Marine products; fresh, chilled, frozen, salted, dehydrated, eviscerated marine produce which can be recognized its characteristics	NFIS (National Fisheries Products Quality Inspection Service)

National Institute of Food and Drug Safety Evaluation

National Institute of Food and Drug Safety Evaluation (NIFDS) is functioning as well. The National Institute of Food and Drug Safety Evaluation is a national organization for toxicological tests and research. Under the Korea Food & Drug Administration, the Institute performs research on toxicology, pharmacology, and risk analysis of foods, drugs, and their additives. The Institute strives primarily to understand important biological triggering mechanisms and improve assessment methods of human exposure, sensitivities, and risk by (1) conducting basic, applied, and policy research that closely examines biologically triggering harmful effects on the regulated products such as foods, food additives, and drugs, and operating the national toxicology program for the toxicological test development and inspection of hazardous chemical substances assessments. The Institute ensures safety by investigation and research on safety by its own researchers, contract research by external academicians and research centers.

Taiwan

In Taiwan Health and Welfare Ministry in charge of Food and Drug Safety, also evaluate the Catering industry to maintenance the food product quality.

United Kingdom

In the UK the Food Standards Agency is an independent government department responsible for food safety and hygiene across the UK. They work with businesses to help them produce safe food, and with local authorities to enforce food safety regulations. In 2006 food hygiene legislation changed and new requirements came into force. The main requirement resulting from this change is that if you own or run a food business in the UK, you must have a documented Food Safety Management System, which is based on the principles of Hazard Analysis Critical Control Point HACCP.

United States

The US food system is regulated by numerous federal, state and local officials. It has been criticized as lacking in "organization, regulatory tools, and not addressing food borne illness."

Federal Level Regulation

The Food and Drug Administration publishes the Food Code, a model set of guidelines and procedures that assists food control jurisdictions by providing a scientifically sound technical and legal basis for regulating the retail and food service industries, including restaurants, grocery stores and institutional foodservice providers such as nursing homes. Regulatory agencies at all levels of government in the United States use the FDA Food Code to develop or update food safety rules in their jurisdictions that are consistent with national food regulatory policy. According to the FDA, 48 of 56 states and territories, representing 79% of the U.S. population, have adopted food codes patterned after one of the five versions of the Food Code, beginning with the 1993 edition.

In the United States, federal regulations governing food safety are fragmented and complicated, according to a February 2007 report from the Government Accountability Office. There are 15 agencies sharing oversight responsibilities in the food safety system, although the two primary agencies are the U.S. Department of Agriculture (USDA) Food Safety and Inspection Service (FSIS), which is responsible for the safety of meat, poultry, and processed egg products, and the Food and Drug Administration (FDA), which is responsible for virtually all other foods.

The Food Safety and Inspection Service has approximately 7,800 inspection program personnel working in nearly 6,200 federally inspected meat, poultry and processed egg establishments. FSIS is charged with administering and enforcing the Federal Meat Inspection Act, the Poultry Products Inspection Act, the Egg Products Inspection Act, portions of the Agricultural Marketing Act, the Humane Slaughter Act, and the regulations that implement these laws. FSIS inspection program personnel inspect every animal before slaughter, and each carcass after slaughter to ensure public health requirements are met. In fiscal year (FY) 2008, this included about 50 billion pounds of livestock carcasses, about 59 billion pounds of poultry carcasses, and about 4.3 billion pounds of processed egg products. At U.S. borders, they also inspected 3.3 billion pounds of imported meat and poultry products.

Industry Pressure

There have been concerns over the efficacy of safety practices and food industry pressure on U.S. regulators. A study reported by Reuters found that “the food industry is jeopardizing U.S. public health by withholding information from food safety investigators or pressuring regulators to withdraw or alter policy designed to protect consumers”. A survey found that 25% of U.S. government inspectors and scientists surveyed have experienced during the past year corporate interests forcing their food safety agency to withdraw or to modify agency policy or action that protects consumers. Scientists have observed that management undercuts field inspectors who stand up for food safety against industry pressure. According to Dr. Dean Wyatt, a USDA veterinarian who oversees federal slaughter house inspectors, “Upper level management does not adequately support field inspectors and the actions they take to protect the food supply. Not only is there lack of support, but there’s outright obstruction, retaliation and abuse of power.” A growing number of food and beverage manufacturers are improving food safety standards by incorporating a food safety management system which automates all steps in the food quality management process.

State and Local Regulation

A number of U.S. states have their own meat inspection programs that substitute for USDA inspection for meats that are sold only in-state. Certain state programs have been criticized for undue leniency to bad practices.

However, other state food safety programs supplement, rather than replace, Federal inspections, generally with the goal of increasing consumer confidence in the state's produce. For example, state health departments have a role in investigating outbreaks of food-borne disease bacteria, as in the case of the 2006 outbreak of *Escherichia coli* O157:H7 (a pathogenic strain of the ordinarily harmless bacteria, *E. coli*) from processed spinach. Health departments also promote better food processing practices to eliminate these threats.

In addition to the US Food and Drug Administration, several states that are major producers of fresh fruits and vegetables (including California, Arizona and Florida) have their own state programs to test produce for pesticide residues.

Restaurants and other retail food establishments fall under state law and are regulated by state or local health departments. Typically these regulations require official inspections of specific design features, best food-handling practices, and certification of food handlers. In some places a letter grade or numerical score must be prominently posted following each inspection. In some localities, inspection deficiencies and remedial action are posted on the Internet.

Vietnam

The Vietnam Food Administration is responsible for managing food hygiene, safety, and quality and has made significant progress since its establishment in 1999. Food safety remains a high priority in Vietnam with the growth of export markets and increasing food imports raising the need to rapidly build capacity of the Food Administration in order to reduce threats of foodborne disease. The Food Administration has demonstrated commitment to the food safety challenges it faces, and has embarked on an innovative capacity building activity with technical assistance from the World Health Organization.

Consumer Labeling

United Kingdom

Foodstuffs in the UK have one of two labels to indicate the nature of the deterioration of the product and any subsequent health issues. EHO Food Hygiene certification is required to prepare and distribute food. While there is no specified expiry date of such a qualification the changes in legislation it is suggested to update every five years.

Best before indicates a future date beyond which the food product *may* lose quality in terms of taste or texture amongst others, but does not imply any serious health problems if food is consumed beyond this date (within reasonable limits).

Use by indicates a legal date beyond which it is not permissible to sell a food product (usually one that deteriorates fairly rapidly after production) due to the potential serious nature of consumption of pathogens. Leeway is sometimes provided by producers in stating display until dates so that products are not at their limit of safe consumption on the actual date stated (this latter is voluntary and not subject to regulatory control). This allows for the variability in production, storage and display methods.

United States

With the exception of infant formula and baby foods which must be withdrawn by their expiration date, Federal law does not require expiration dates. For all other foods, except dairy products in some states, freshness dating is strictly voluntary on the part of manufacturers. In response to consumer demand, perishable foods are typically labelled with a Sell by date. It is up to the consumer to decide how long after the Sell by date a package is usable. Other common dating statements are Best if used by, Use-by date, Expiration date, Guaranteed fresh <date>, and Pack date.

Australia and New Zealand

Guide to Food Labelling and Other Information Requirements: This guide provides background information on the general labelling requirements in the Code. The information in this guide applies both to food for retail sale and to food for catering purposes. Foods for catering purposes means those foods for use in restaurants, canteens, schools, caterers or self-catering institutions, where food is offered for immediate consumption. Labelling and information requirements in the new Code apply both to food sold or prepared for sale in Australia and New Zealand and food imported into Australia and New Zealand. Warning and Advisory Declarations, Ingredient Labelling, Date Marking, Nutrition Information Requirements, Legibility Requirements for Food Labels, Percentage Labelling, Information Requirements for Foods Exempt from Bearing a Label.

Food Quality

Food quality is the quality characteristics of food that is acceptable to consumers. This includes external factors as appearance (size, shape, color, gloss, and consistency), texture, and flavor; factors such as federal grade standards (e.g. of eggs) and internal (chemical, physical, microbial).

[[Food quality in the United States is enforced by the Food Safety Act 1990. Members of the public complain to trading standards professionals[specify], who submit complaint samples and also samples used to routinely monitor the food marketplace to public analysts. Public analysts carry out scientific analysis on the samples to determine whether the quality is of sufficient standard.]]

Food quality is an important food manufacturing requirement, because food consumers are susceptible to any form of contamination that may occur during the manufacturing process. Many consumers also rely on manufacturing and processing standards, particularly to know what ingredients are present, due to dietary, nutritional requirements (kosher, halal, vegetarian), or medical conditions (e.g., diabetes, or allergies).

Besides ingredient quality, there are also sanitation requirements. It is important to ensure that the food processing environment is as clean as possible in order to produce the safest possible food for the consumer. A recent example of poor sanitation recently has been the 2006 North American E. coli outbreak involving spinach, an outbreak that is still under investigation after new information has come to light regarding the involvement of Cambodian nationals.

Food quality also deals with product traceability, (e.g., of ingredient, and packaging suppliers), should a recall of the product be required. It also deals with labeling issues to ensure there is correct ingredient and nutritional information.

There are many existing international quality institutes testing food products in order to indicate to all consumers which are higher quality products. Founded in 1961 in Brussels, The international quality institute Monde Selection is the oldest one in evaluating food quality. During the degustations, the products must meet the following selection criteria, required by the Institute: sensory analysis, bacteriological and chemical analysis, the nutrition and health claims, and the utilization notice. In short, the judgments are based on the following areas: taste, health, convenience, labeling, packaging, environmental friendliness and innovation. As many consumers rely on manufacturing and processing standards, the Institute Monde Selection takes into account the European Food Law.

References

- "Industry Has Sway Over Food Safety System: U.S. Study". Reuters. 2010-09-14. Archived from the original on 2010-09-14.
- "Warning and Advisory Statements and Declarations User Guide to Standard 1.2.3 – Mandatory Warning and Advisory Statements and Declarations" (PDF). Food Standards Australia New Zealand. Retrieved 2010-09-15.

2

Methods for Food Safety

The measures taken for food safety are the preservation of food, storage of food and food defense. While food defense is the protection of food products from adulteration by biological, chemical and physical agents, food preservation involves the prevention of bacteria, fungi and also slows the oxidation of fats. This chapter helps develop a comprehensive understanding on food safety.

Food Storage

Yup'ik elevated food cache (*qulvarvik*), Hooper Bay, Alaska, 1929. Photograph by Edward S. Curtis

Food logistics is both a traditional domestic skill and is important industrially. The transport and storing of food as well as timely delivery to consumers is important in order to secure the procurement of food. Food is stored by almost every human society and by many animals. Storing of food has several main purposes:

- Storage of harvested and processed plant and animal food products for distribution to consumers
- Enabling a better balanced diet throughout the year
- Reducing kitchen waste by preserving unused or uneaten food for later use
- Preserving pantry food, such as spices or dry ingredients like rice and flour, for eventual use in cooking
- Preparedness for catastrophes, emergencies and periods of food scarcity or famine

- Religious reasons (Example: LDS Church leaders instruct church members to store food)
- Protection from animals or theft

U.S. Federal Emergency Management Agency (FEMA) food storage containers stacked on shipping pallets in Texas, 2008.

History of Food Storage

Traditional Sami food storage structure to protect food from animals

Domestic Food Storage

Tupperware kitchen storage containers designed for a variety of uses.

The safe storage of food for home use should strictly adhere to guidelines set out by reliable sources, such as the United States Department of Agriculture. These guidelines have been thoroughly researched by scientists to determine the best methods for reducing the real threat of food poisoning from unsafe food storage. It is also important to maintain proper kitchen hygiene, to reduce risks of bacteria or virus growth and food poisoning. The common food poisoning illnesses include

Listeriosis, Mycotoxicosis, Salmonellosis, E. coli, Staphylococcal food poisoning and Botulism. There are many other organisms that can also cause food poisoning.

Plastic storage containers can be used to store food.

There are also safety guidelines available for the correct methods of home canning of food. For example, there are specific boiling times that apply depending upon whether pressure canning or waterbath canning is being used in the process. These safety guidelines are intended to reduce the growth of mold and bacteria and the threat of potentially-fatal food poisoning.

Food Storage Safety

Freezers and Thawing Food

Freezer temperature should be maintained below 0°F. Food should never be thawed at room temperature, this increases the risk of bacterial and fungal growth and accordingly the risk of food poisoning. Once thawed, food should be used and never refrozen. Frozen food should be thawed using the following methods:

- Microwave oven
- During cooking
- In cold water (place food in watertight, plastic bag; change water every 30 minutes)
- In the refrigerator

Throw out foods that have been warmer than 40 °F for more than 2 hours. If there is any doubt at all about the length of time the food has been defrosted at room temperature, it should be thrown out. Freezing does not destroy microbes present in food. Freezing at 0 °F does inactivate microbes (bacteria, yeasts and molds). However, once food has been thawed, these microbes can again become active. Microbes in thawed food can multiply to levels that can lead to foodborne illness. Thawed food should be handled according to the same guidelines as perishable fresh food.

Food frozen at 0°F and below is preserved indefinitely. However, the quality of the food will deteriorate if it is frozen over a lengthy period. The United States Department of Agriculture, Food Safety and Inspection Service publishes a chart showing the suggested freezer storage time for common foods.

Refrigeration

It is important to note that safe food storage using refrigeration requires adhering to temperature guidelines:

For safety, it is important to verify the temperature of the refrigerator. Refrigerators should be set to maintain a temperature of 40 °F or below. Some refrigerators have built-in thermometers to measure their internal temperature. For those refrigerators without this feature, keep an appliance thermometer in the refrigerator to monitor the temperature. This can be critical in the event of a power outage. When the power goes back on, if the refrigerator is still 40 °F, the food is safe. Foods held at temperatures above 40 °F for more than 2 hours should not be consumed. Appliance thermometers are specifically designed to provide accuracy at cold temperatures. Be sure refrigerator/ freezer doors are closed tightly at all times. Don't open refrigerator/freezer doors more often than necessary and close them as soon as possible.

Storage Times for Refrigerated Food

The United States Department of Agriculture, Food Safety and Inspection Service publishes recommended storage times for refrigerated food.

Storing Oils and Fats

Oils and fats can begin to go rancid quickly when not stored safely. Rancid cooking oils and fats do not often smell rancid until well after they have spoiled. Oxygen, light and heat all contribute to cooking oils becoming rancid. The higher the level of polyunsaturated fat that an oil contains, the faster it spoils. The percentage of polyunsaturated fat in some common cooking oils is: safflower (74%); sunflower (66%); corn (60%); soybean (37%); peanut (32%); canola (29%); olive (8%).

To help preserve oils from rancidification, they should be refrigerated once opened. Opened, refrigerated cooking oils should be used within a few weeks, when some types begin to go rancid. Unopened oils can have a storage life of up to one year, but some types have a shorter shelf-life even when unopened (such as sesame and flaxseed).

Dry Storage of Foods

Vegetables

A large root cellar at the Oxon Hill Manor farm in Maryland

The guidelines vary for safe storage of vegetables under dry conditions (without refrigerating or freezing). This is because different vegetables have different characteristics, for example, tomatoes contain a lot of water, while root vegetables such as carrots and potatoes contain less. These factors, and many others, affect the amount of time that a vegetable can be kept in dry storage, as well as the temperature needed to preserve its usefulness. The following guideline shows the required dry storage conditions:

- Cool and dry: onion
- Cool and moist: root vegetable, potato, cabbage
- Warm and dry: winter squash, pumpkin, sweet potatoes, dried hot peppers

Many cultures have developed innovative ways of preserving vegetables so that they can be stored for several months between harvest seasons. Techniques include pickling, home canning, food dehydration, or storage in a root cellar.

Grain

Grain, which includes dry kitchen ingredients such as flour, rice, millet, couscous, cornmeal, and so on, can be stored in rigid sealed containers to prevent moisture contamination or insect or rodent infestation. For kitchen use, glass containers are the most traditional method. During the 20th century plastic containers were introduced for kitchen use. They are now sold in a vast variety of sizes and designs.

Metal cans are used (in the United States the smallest practical grain storage uses closed-top #10 metal cans). Storage in grain sacks is ineffective; mold and pests destroy a 25 kg cloth sack of grain in a year, even if stored off the ground in a dry area. On the ground or damp concrete, grain can spoil in as little as three days, and the grain might have to be dried before it can be milled. Food stored under unsuitable conditions should not be purchased or used because of risk of spoilage. To test whether grain is still good, it can be sprouted. If it sprouts, it is still good, but if not, it should not be eaten. It may take up to a week for grains to sprout. When in doubt about the safety of the food, throw it out as quickly as possible.

Spices and Herbs

Spices and herbs are today often sold prepackaged in a way that is convenient for pantry storage. The packaging has dual purposes of both storing and dispensing the spices or herbs. They are sold in small glass or plastic containers or resealable plastic packaging. When spices or herbs are homegrown or bought in bulk, they can be stored at home in glass or plastic containers. They can be stored for extended periods, in some cases for years. However, after 6 months to a year, spices and herbs will gradually lose their flavour as oils they contain will slowly evaporate during storage.

Spices and herbs can be preserved in vinegar for short periods of up to a month, creating a flavoured vinegar.

Alternative methods for preserving herbs include freezing in water or unsalted butter. Herbs can be chopped and added to water in an ice cube tray. After freezing, the ice cubes are emptied into a plastic freezer bag for storing in the freezer. Herbs also can be stirred into a bowl with unsalted butter, then spread on wax paper and rolled into a cylinder shape. The wax paper roll containing the butter and herbs is then stored in a freezer, and can be cut off in the desired amount for cooking. Using either of these techniques, the herbs should be used within a year.

Meat

Unpreserved meat has only a relatively short life in storage. Perishable meats should be refrigerated, frozen, dried promptly or cured. Storage of fresh meats is a complex discipline that affects the costs, storage life and eating quality of the meat, and the appropriate techniques vary with the kind of meat and the particular requirements. For example, dry ageing techniques are sometimes used to tenderize gourmet meats by hanging them in carefully controlled environments for up to 21 days, while game animals of various kinds may be hung after shooting. Details depend on personal tastes and local traditions. Modern techniques of preparing meat for storage vary with the type of meat and special requirements of tenderness, flavour, hygiene, and economy.

Semi-dried meats like salamis and country style hams are processed first with salt, smoke, sugar, acid, or other "cures" then hung in cool dry storage for extended periods, sometimes exceeding a year. Some of the materials added during the curing of meats serve to reduce the risks of food poisoning from anaerobic bacteria such as species of Clostridium that release botulinum toxin that can cause botulism. Typical ingredients of curing agents that inhibit anaerobic bacteria include nitrates. Such salts are dangerously poisonous in their own right and must be added in carefully controlled quantities and according to proper techniques. Their proper use has however saved many lives and much food spoilage.

Like the semi-dried meats, most salted, smoked, and simply-dried meats of different kinds that once were staples in particular regions, now are largely luxury snacks or garnishes; examples include jerky, biltong, and varieties of pemmican, but ham and bacon for instance, still are staples in many communities.

Food Rotation

Food rotation is important to preserve freshness. When food is rotated, the food that has been in storage the longest is used first. As food is used, new food is added to the pantry to replace it; the essential rationale is to use the oldest food as soon as possible so that nothing is in storage too long and becomes unsafe to eat. Labelling food with paper labels on the storage container, marking the date that the container is placed in storage, can make this practice simpler. The best way to rotate food storage is to prepare meals with stored food on a daily basis.

For Emergency Preparation

Guides for surviving emergency conditions in many parts of the world recommend maintaining a store of essential foods; typically water, cereals, oil, dried milk, and protein rich foods such as beans, lentils, tinned meat and fish. A food storage calculator can be used to help determine how much of these staple foods a person would need to store in order to sustain life for one full year. In addition to storing the basic food items many people choose to supplement their food storage with frozen or preserved garden-grown fruits and vegetables and freeze-dried or canned produce. An unvarying diet of staple foods prepared in the same manner can cause appetite exhaustion, leading to less caloric intake. Another benefit to having a basic supply of food storage in the home is for the potential cost savings. Costs of dry bulk foods (before preparation) are often considerably less than convenience and fresh foods purchased at local markets or supermarkets. There is a significant market in convenience foods for campers, such as dehydrated food products.

Commercial Food Logistics

Silos connected to a grain elevator on a farm in Israel.

Grain and beans are stored in tall grain elevators, almost always at a rail head near the point of production. The grain is shipped to a final user in hopper cars. In the former Soviet Union, where harvest was poorly controlled, grain was often irradiated at the point of production to suppress mold and insects. In the U.S., threshing and drying is performed in the field, and transport is nearly sterile and in large containers that effectively suppresses pest access, which eliminates the need for irradiation. At any given time, the U.S. usually has about two weeks worth of stored grains for the population.

Fresh fruits and vegetables are sometimes packed in plastic packages and cups for fresh premium markets, or placed in large plastic tubs for sauce and soup processors. Fruits and vegetables are usually refrigerated at the earliest possible moment, and even so have a shelf life of two weeks or less.

In the United States, livestock is usually transported live, slaughtered at a major distribution point, hung and transported for two days to a week in refrigerated rail cars, and then butchered and sold locally. Before refrigerated rail cars, meat had to be transported live, and this placed its cost so high that only farmers and the wealthy could afford it every day. In Europe much meat is transported live and slaughtered close to the point of sale. In much of Africa and Asia most meat is for local populations is raised, slaughtered and eaten locally, which is believed to be less stressful for the animals involved and minimizes meat storage needs. In Australia and New Zealand, where a large proportion of meat production is for export, meat enters the cold chain early, being stored in large freezer plants before being shipped overseas in freezer ships.

Food Defense

Food defense is the protection of food products from intentional contamination or adulteration by biological, chemical, physical, or radiological agents. It addresses additional concerns including physical, personnel and operational security. This is in contrast to food safety, which is based on accidental or environmental contamination, and food security, which deals with individuals having access to enough food for an active, healthy life. These three terms are often conflated. Food protection is the umbrella term encompassing both food defense and food safety.

Along with protecting the food system, food defense also deals with prevention, protection, mitigation, response and recovery from intentional acts of adulteration.

History in the USA

- 1906: The Federal Meat Inspection Act places requirements on the slaughter, processing and labeling of meat and meat products, domestic and imported.
- 1938: The Federal Food, Drug and Cosmetic Act establishes definitions and regulation for the safety of food, drugs, and cosmetics.
- 1957: The Poultry Products Inspection Act requires the Food Safety and Inspection Service (FSIS) to inspect all domesticated birds meant for human consumption.
- November 2002: The Homeland Security Act passed by congress creates the Department of Homeland Security in response to the September 11 attacks.
- December 2003: Homeland Security Presidential Directive 7 establishes a policy to identify and prioritize critical infrastructures. Food and Agriculture is identified as one of these infrastructures
- January 2004: The Homeland Security Presidential Directive 9 establishes policy to protect agriculture and food systems against terrorist attacks.
- January 2004: DHS launches Homeland Security Centers of Excellence in order to conduct research to address homeland security challenges.
- July 2004: The National Center for Food Protection and Defense now Food Protection and Defense Institute (FPDI) is formally launched with the vision to "defend the safety of the food system through research and education" at the University of Minnesota, Twin Cities.
- January 2011: The Food Safety Modernization Act grants the FDA new authorities and powers, as well as directly providing for protections against intentional adulteration.

Food Defense Event Types

Food defense events can generally be categorized into three types. These could be carried out by a disgruntled employee, sophisticated insider, or intelligent adversary with a specific goal in mind. This goal may be to impact the public, brand, company or the psycho-social stability of a group of people depending on the type. However an event may contain aspects of more than one category.

Industrial Sabotage

These events include intentional contamination by a disgruntled employee, insider or competitor with the intention of damaging the brand of the company, causing financial problems from a widespread recall or sabotage, but not necessarily with the goal of causing widespread illness or public harm. These internal actors often know what procedures are followed in the plant, and how to bypass checkpoints and security controls.

An example of a disgruntled employee is the contamination of frozen foods produced by a subsidiary of Maruha Nichiro Holdings with malathion, a pesticide. The contamination is speculated to have occurred because the employee was dissatisfied with pay and benefits. The contamination resulted in a recall of 6.4 million potentially tainted products. Nearly 1,800 people are estimated to have been affected, and public confidence in food quality was shaken.

Terrorism

The reach and complexity of the food system has caused concern for its potential as a terrorist target.

The first and largest food attack in the US is the 1984 Rajneeshee bioterror attack. 751 individuals were poisoned in The Dalles, Oregon through the contamination of salad bars with Salmonella with the intention of affecting the 1984 Wasco County elections.

Economically Motivated Adulteration (EMA)

The FDA's working definition of EMA is the fraudulent, intentional substitution or addition of a substance in a product for the purpose of increasing the apparent value of the product or reducing the cost of its production, i.e., for economic gain. EMA includes dilution of products with increased quantities of an already-present substance (e.g., increasing inactive ingredients of a drug with a resulting reduction in strength of the finished product, or watering down of juice) to the extent that such dilution poses a known or possible health risk to consumers, as well as the addition or substitution of substances in order to mask dilution.

EMA, also known as Food Fraud, commonly occurs for financial advantage through the undeclared substitution with alternative ingredients. This poses a health concern due to allergen labeling requirements. These deliberate acts are intended to evade detection, posing a challenge to regulating bodies and quality assurance methodologies.

Cases of EMA have been seen in the horse-meat scandal, melamine contamination scandal and the *Salmonella* outbreak involving the Peanut Corporation of America. The most commonly counterfeited product is extra-virgin olive oil. Other products commonly associated with food fraud include fish and seafood, honey, meat and grain-based foods, fruit juices, organic foods, coffee, some highly processed foods, tea and spices. Experts estimate that up to 10% of food products in retail stores contain some degree of adulteration, and EMA events cost the US food industry between $10 billion and $15 billion a year.

Protection Strategies

Regulatory bodies and industry can implement strategies and use tools in order to protect their supply chains and processing facilities from intentional contamination or adulteration. Defined as protection or mitigation, this process involves both assessing the risk and vulnerabilities of a single supply chain or facility and working to mitigate these risks in order to prevent an event, and reducing the severity of an event.

Tools

The FDA has developed several tools for the food industry, including among others:

- Mitigation Strategies Database which includes a range of preventative measures and suggestions for companies
- Food Defense 101, focused on training for preparedness against an intentional attack
- FREE-B, a compilation of both intentional and unintentional food contamination scenarios
- Food Defense Plan Builder, designed to assist owners and operators of food facilities with developing personalized food defense plans

Risk Assessments

It is difficult to quantify the risk in a system, due to the stochastic nature of the events. However, it is possible to use other sources of information, such as gathered intelligence, economic and social drivers and data mining to assess the potential weaknesses and entry points of a system, along with the scale of consequences related to a breach in that system. Tools being developed for this purpose by the National Center for Food Protection and Defense include Focused Integration of Data for Early Signals (FIDES) and Criticality Spatial Analysis (CRISTAL).

Food industry stakeholders can perform a vulnerability assessment to understand the vulnerabilities of their system, the consequences of an event and the potential threats and agents. This allows companies to assess and prioritize vulnerabilities within their facility and system. A software tool has been developed by the FDA to assist with this process. Companies are encouraged to create a Food Defense Plan based on the vulnerability and risk assessments performed, detailing their plan of action in the case of an intentional or unintentional contamination event.

Vulnerability Assessment Tools

The FDA has identified four key activities, or common vulnerabilities within the food system: bulk liquid receiving and loading, liquid storage and handling, secondary ingredient handling, and mixing or similar activities. Knowledge of these key activities can direct action plans.

CARVER + Shock is used to consider the factors involved in an intentional contamination event

- Criticality - measure of public health and economic impacts of an attack
- Accessibility – ability to physically access and egress from target
- Recuperability – ability of system to recover from an attack
- Vulnerability – ease of accomplishing attack

- Effect – amount of direct loss from an attack as measured by loss in production
- Recognizability – ease of identifying target
- Shock – the combined health, economic, and psychological impacts of an attack

Supply Chain Control

Understanding the supply chains involved in a food system is difficult due to their complex and often obscured nature, but having a good understanding of where incoming ingredients come from can help to mitigate contamination and adulteration. Good supply chain management coupled with regular audits and quality assurance analyses can help to safeguard companies from contamination originating outside the facility.

In addition, companies can take advantage of existing scenario based tools and should follow Good Manufacturing Practice guidelines.

Mitigation Strategies

Physical Measures

- Secure the facility perimeter and perform periodic checks
- Use controlled-access procedures for people or vehicles entering the plant or parking area
- Install an alarm system, cameras and sufficient lighting
- Designate restricted areas for authorized employees, restrict non-employees to non-production areas
- Limit access to control systems
- Use tamper-evident or tamper-resistant packaging

Policy Measures

- Use a system to identify personnel by their specific functions
- Conduct background checks on all employees and contractors who will be working in sensitive operations
- Train employees on food defense and security awareness, including recognition of suspicious behavior or individuals

Management

- Maintain records to allow easy trace-back and trace-forward of materials and products
- Have available a list of contacts for local, state and federal agencies
- Implement an inventory control system

More strategies for various categories and nodes of the food system can be found in the various mitigation strategies databases available through the FDA and USDA.

Food Preservation

Food preservation involves preventing the growth of bacteria, fungi (such as yeasts), or other micro-organisms (although some methods work by introducing benign bacteria or fungi to the food), as well as slowing the oxidation of fats that cause rancidity. Food preservation may also include processes that inhibit visual deterioration, such as the enzymatic browning reaction in apples after they are cut during food preparation.

Many processes designed to preserve food will involve a number of food preservation methods. Preserving fruit by turning it into jam, for example, involves boiling (to reduce the fruit's moisture content and to kill bacteria, etc.), sugaring (to prevent their re-growth) and sealing within an airtight jar (to prevent recontamination). Some traditional methods of preserving food have been shown to have a lower energy input and carbon footprint, when compared to modern methods. However, some methods of food preservation are known to create carcinogens, and in 2015, the International Agency for Research on Cancer of the World Health Organization classified processed meat, i.e. meat that has undergone salting, curing, fermenting, and smoking, as "carcinogenic to humans".

Maintaining or creating nutritional value, texture and flavor is an important aspect of food preservation, although, historically, some methods drastically altered the character of the food being preserved. In many cases these changes have come to be seen as desirable qualities – cheese, yogurt and pickled onions being common examples.

Traditional Techniques

New techniques of food preservation became available to the home chef from the dawn of agriculture until the Industrial Revolution.

Drying

Drying is one of the oldest techniques used to hamper the decomposition of food products. As early as 12,000 B.C., Middle Eastern and Oriental cultures were drying foods using the power of the sun. Vegetables and fruit are naturally dried by the sun and wind, but "still houses" were built in areas that did not have enough sunlight to dry things. A fire would be built inside the building to provide the heat to dry the various fruits, vegetables, and herbs.

Cooling

Cooling preserves foods by slowing down the growth and reproduction of micro-organisms and the action of enzymes that cause food to rot. The introduction of commercial and domestic refrigerators drastically improved the diets of many in the Western world by allowing foods such as fresh fruit, salads and dairy products to be stored safely for longer periods, particularly during warm weather.

Freezing

Freezing is also one of the most commonly used processes, both commercially and domestically, for preserving a very wide range of foods, including prepared foods that would not have required freezing in their unprepared state. For example, potato waffles are stored in the freezer, but potatoes themselves require only a cool dark place to ensure many months' storage. Cold stores provide large-volume, long-term storage for strategic food stocks held in case of national emergency in many countries.

Boiling

Boiling liquid food items can kill any existing microbes. Milk and water are often boiled to kill any harmful microbes that may be present in them.

Heating

Heating to temperatures which are sufficient to kill microorganisms inside the food is a method used with perpetual stews. Milk is also boiled before storing to kill many microorganisms.

Salting

Bag of Prague powder #1, also known as "curing salt" or "pink salt". It is typically a combination of salt and sodium nitrite, with the pink color added to distinguish it from ordinary salt.

Salting or curing draws moisture from the meat through a process of osmosis. Meat is cured with salt or sugar, or a combination of the two. Nitrates and nitrites are also often used to cure meat and contribute the characteristic pink color, as well as inhibition of *Clostridium botulinum*. It was a main method of preservation in medieval times and around the 1700s.

Sugaring

The earliest cultures have used sugar as a preservative, and it was commonplace to store fruit in honey. Similar to pickled foods, sugar cane was brought to Europe through the trade routes. In northern climates without sufficient sun to dry foods, preserves are made by heating the fruit with sugar. "Sugar tends to draw water from the microbes (plasmolysis). This process leaves the microbial cells dehydrated, thus killing them. In this way, the food will remain safe from microbial spoilage." Sugar is used to preserve fruits, either in an anti-microbial syrup with fruit such as apples, pears, peaches, apricots and plums, or in crystallized form where the preserved material is cooked in sugar to the point of crystallization and the resultant product is then stored dry. This method is used for the skins of citrus fruit (candied peel), angelica and ginger. Also sugaring can be used in jam jellies.

Smoking

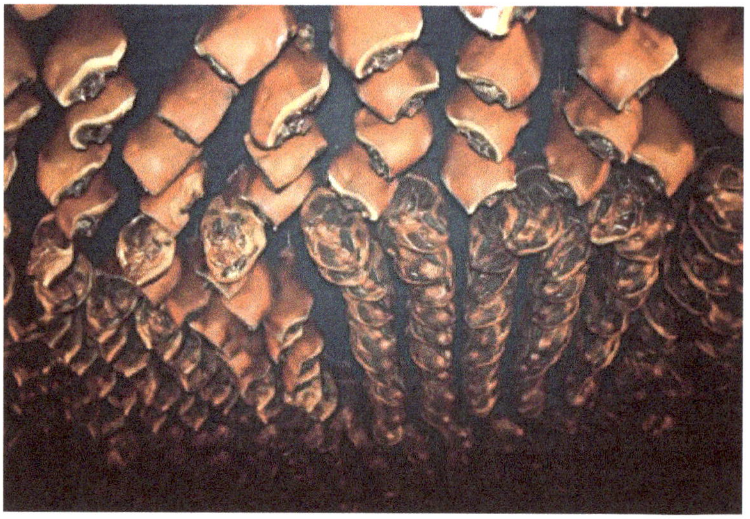

Smoked ham in a smokehouse in Schleswig-Holstein, Germany

Smoking is used to lengthen the shelf life of perishable food items. This effect is achieved by exposing the food to smoke from burning plant materials such as wood. Smoke deposits a number of pyrolysis products onto the food, including the phenols syringol, guaiacol and catechol. These compounds aid in the drying and preservation of meats and other foods. Most commonly subjected to this method of food preservation are meats and fish that have undergone curing. Fruits and vegetables like paprika, cheeses, spices, and ingredients for making drinks such as malt and tea leaves are also smoked, but mainly for cooking or flavoring them. It is one of the oldest food preservation methods, which probably arose after the development of cooking with fire.

Pickling

Pickling is a method of preserving food in an edible anti-microbial liquid. Pickling can be broadly classified into two categories: chemical pickling and fermentation pickling.

In chemical pickling, the food is placed in an edible liquid that inhibits or kills bacteria and other micro-organisms. Typical pickling agents include brine (high in salt), vinegar, alcohol, and vegetable oil, especially olive oil but also many other oils. Many chemical pickling processes also involve heating or boiling so that the food being preserved becomes saturated with the pickling agent. Common chemically pickled foods include cucumbers, peppers, corned beef, herring, and eggs, as well as mixed vegetables such as piccalilli.

In fermentation pickling, the food itself produces the preservation agent, typically by a process that produces lactic acid. Fermented pickles include sauerkraut, nukazuke, kimchi, surströmming, and curtido. Some pickled cucumbers are also fermented.

Lye

Sodium hydroxide (lye) makes food too alkaline for bacterial growth. Lye will saponify fats in the food, which will change its flavor and texture. Lutefisk uses lye in its preparation, as do some olive recipes. Modern recipes for century eggs also call for lye.

Canning

Preserved food

Canning involves cooking food, sealing it in sterile cans or jars, and boiling the containers to kill or weaken any remaining bacteria as a form of sterilization. It was invented by the French confectioner Nicolas Appert. By 1806, this process was used by the French Navy to preserve meat, fruit, vegetables, and even milk. Although Appert had discovered a new way of preservation, it wasn't understood until 1864 when Louis Pasteur found the relationship between microorganisms, food spoilage, and illness.

Foods have varying degrees of natural protection against spoilage and may require that the final step occur in a pressure cooker. High-acid fruits like strawberries require no preservatives to can and only a short boiling cycle, whereas marginal vegetables such as carrots require longer boiling and addition of other acidic elements. Low-acid foods, such as vegetables and meats, require pressure canning. Food preserved by canning or bottling is at immediate risk of spoilage once the can or bottle has been opened.

Lack of quality control in the canning process may allow ingress of water or micro-organisms. Most such failures are rapidly detected as decomposition within the can causes gas production and the can will swell or burst. However, there have been examples of poor manufacture (under-processing) and poor hygiene allowing contamination of canned food by the obligate anaerobe *Clostridium botulinum*, which produces an acute toxin within the food, leading to severe illness or death. This organism produces no gas or obvious taste and remains undetected by taste or smell. Its toxin is denatured by cooking, however. Cooked mushrooms, handled poorly and then canned, can support the growth of Staphylococcus aureus, which produces a toxin that is not destroyed by canning or subsequent reheating.

Jellying

Food may be preserved by cooking in a material that solidifies to form a gel. Such materials include gelatin, agar, maize flour, and arrowroot flour. Some foods naturally form a protein gel when cooked, such as eels and elvers, and sipunculid worms, which are a delicacy in Xiamen, in the Fujian province of the People's Republic of China. Jellied eels are a delicacy in the East End of London, where they are eaten with mashed potatoes. Potted meats in aspic (a gel made from gelatine and clarified meat broth) were a common way of serving meat off-cuts in the UK until the 1950s. Many jugged meats are also jellied.

A traditional British way of preserving meat (particularly shrimp) is by setting it in a pot and sealing it with a layer of fat. Also common is potted chicken liver; compare pâté.

Jugging

Meat can be preserved by jugging. Jugging is the process of stewing the meat (commonly game or fish) in a covered earthenware jug or casserole. The animal to be jugged is usually cut into pieces, placed into a tightly-sealed jug with brine or gravy, and stewed. Red wine and/or the animal's own blood is sometimes added to the cooking liquid. Jugging was a popular method of preserving meat up until the middle of the 20th century.

Burial

Burial of food can preserve it due to a variety of factors: lack of light, lack of oxygen, cool temperatures, pH level, or desiccants in the soil. Burial may be combined with other methods such as salting or fermentation. Most foods can be preserved in soil that is very dry and salty (thus a desiccant) such as sand, or soil that is frozen.

Many root vegetables are very resistant to spoilage and require no other preservation than storage in cool dark conditions, for example by burial in the ground, such as in a storage clamp. Century eggs are created by placing eggs in alkaline mud (or other alkaline substance), resulting in their "inorganic" fermentation through raised pH instead of spoiling. The fermentation preserves them and breaks down some of the complex, less flavorful proteins and fats into simpler, more flavorful ones. Cabbage was traditionally buried in the fall in northern farms in the U.S. for preservation. Some methods keep it crispy while other methods produce sauerkraut. A similar process is used in the traditional production of kimchi. Sometimes meat is buried under conditions that cause preservation. If buried on hot coals or ashes, the heat can kill pathogens, the dry ash can desiccate, and the earth can block oxygen and further contamination. If buried where the earth is very cold, the earth acts like a refrigerator.

In Orissa, India, it is practical to store rice by burying it underground. This method helps to store for three to six months during the dry season.

Curing

The earliest form of curing was dehydration. To accelerate this process, salt is usually added. In the culinary world, it was common to choose raw salts from various sources (rock salt, sea salt, etc.).

More modern "examples of salts that are used as preservatives include sodium chloride (NaCl), sodium nitrate ($NaNO_3$) and sodium nitrite ($NaNO_2$). Even at mild concentrations (up to 2%), sodium chloride, found in many food products, is capable of neutralizing the antimicrobial character of natural compounds."

Fermentation

Some foods, such as many cheeses, wines, and beers, use specific micro-organisms that combat spoilage from other less-benign organisms. These micro-organisms keep pathogens in check by creating an environment toxic for themselves and other micro-organisms by producing acid or alcohol. Methods of fermentation include, but are not limited to, starter micro-organisms, salt, hops, controlled (usually cool) temperatures and controlled (usually low) levels of oxygen. These methods are used to create the specific controlled conditions that will support the desirable organisms that produce food fit for human consumption.

Fermentation is the microbial conversion of starch and sugars into alcohol. Not only can fermentation produce alcohol, but it can also be a valuable preservation technique. Fermentation can also make foods more nutritious and palatable. For example, drinking water in the Middle Ages was dangerous because it often contained pathogens that could spread disease. When the water is made into beer, the boiling during the brewing process kills any bacteria in the water that could make people sick. Additionally, the water now has the nutrients from the barley and other ingredients, and the microorganisms can also produce vitamins as they ferment.

Industrial/Modern Techniques

Techniques of food preservation were developed in research laboratories for commercial applications.

Pasteurization

Pasteurization is a process for preservation of liquid food. It was originally applied to combat the souring of young local wines. Today, the process is mainly applied to dairy products. In this method, milk is heated at about 70 °C for 15 to 30 seconds to kill the bacteria present in it and cooling it quickly to 10 °C to prevent the remaining bacteria from growing. The milk is then stored in sterilized bottles or pouches in cold places. This method was invented by Louis Pasteur, a French chemist, in 1862.

Vacuum Packing

Vacuum-packing stores food in a vacuum environment, usually in an air-tight bag or bottle. The vacuum environment strips bacteria of oxygen needed for survival. Vacuum-packing is commonly used for storing nuts to reduce loss of flavor from oxidization. A major drawback to vacuum packaging, at the consumer level, is that vacuum sealing can deform contents and rob certain foods, such as cheese, of its flavor.

Artificial Food Additives

Preservative food additives can be *antimicrobial*, which inhibit the growth of bacteria or fungi, including mold, or *antioxidant*, such as oxygen absorbers, which inhibit the oxidation of food constituents. Common antimicrobial preservatives include calcium propionate, sodium nitrate, sodium nitrite, sulfites (sulfur dioxide, sodium bisulfite, potassium hydrogen sulfite, etc.) and disodium EDTA. Antioxidants include BHA and BHT. Other preservatives include formaldehyde (usually in solution), glutaraldehyde (kills insects), ethanol, and methylchloroisothiazolinone.

Irradiation

Irradiation of food is the exposure of food to ionizing radiation. The two types of ionizing radiation used are beta particles (high-energy electrons) and gamma rays (emitted from radioactive sources as cobalt-60 or cesium-137). Treatment effects include killing bacteria, molds, and insect pests, reducing the ripening and spoiling of fruits, and at higher doses inducing sterility. The technology may be compared to pasteurization; it is sometimes called "cold pasteurization", as the product is not heated.

The irradiation process is not directly related to nuclear energy, but does use radioactive isotopes produced in nuclear reactors. Cobalt-60, for example does not occur naturally and can only be produced through neutron bombardment of cobalt-59. Ionizing radiation at high energy levels is hazardous to life (hence its usefulness in sterilisation); for this reason, irradiation facilities have a heavily shielded irradiation room where the process takes place. Radiation safety procedures are used to ensure that neither the workers in such facilities nor the environment receives any radiation dose above administrative limits. Irradiated food does not and cannot become radioactive, and national and international expert bodies have declared food irradiation as wholesome. However, the wholesomeness of consuming such food is disputed by opponents and consumer organizations. National and international expert bodies have declared food irradiation as "wholesome"; organizations of the United Nations, such as the World Health Organization and Food and Agriculture Organization, endorse food irradiation. International legislation on whether food may be irradiated or not varies worldwide from no regulation to full banning. Irradiation may allow lower-quality or contaminated foods to be rendered marketable.

Approximately 500,000 tons of food items are irradiated per year worldwide in over 40 countries. These are mainly spices and condiments with an increasing segment of fresh fruit irradiated for fruit fly quarantine.

Pulsed Electric Field Electroporation

Pulsed electric field (PEF) electroporation is a method for processing cells by means of brief pulses of a strong electric field. PEF holds potential as a type of low-temperature alternative pasteurization process for sterilizing food products. In PEF processing, a substance is placed between two electrodes, then the pulsed electric field is applied. The electric field enlarges the pores of the cell membranes, which kills the cells and releases their contents. PEF for food processing is a developing technology still being researched. There have been limited industrial applications of PEF processing for the pasteurization of fruit juices. To date, several PEF treated juices are available on the market in Europe. Furthermore, for several years a juice pasteurization application in the US has used PEF. For cell disintegration purposes especially potato processors show great interest in PEF technology as an efficient alternative for their preheaters. Potato applications are already operational in the US and Canada. There are also commercial PEF potato applications in various countries in Europe, as well as in Australia, India and China.

Modified Atmosphere

Modifying atmosphere is a way to preserve food by operating on the atmosphere around it. Salad crops that are notoriously difficult to preserve are now being packaged in sealed bags with an atmosphere modified to reduce the oxygen (O_2) concentration and increase the carbon dioxide (CO_2) concentration. There is concern that, although salad vegetables retain their appearance and texture in such conditions, this method of preservation may not retain nutrients, especially vitamins. There are two methods for preserving grains with carbon dioxide. One method is placing a block of dry ice in the bottom and filling the can with the grain. Another method is purging the container from the bottom by gaseous carbon dioxide from a cylinder or bulk supply vessel.

Carbon dioxide prevents insects and, depending on concentration, mold and oxidation from damaging the grain. Grain stored in this way can remain edible for approximately five years.

Nitrogen gas (N_2) at concentrations of 98% or higher is also used effectively to kill insects in the grain through hypoxia. However, carbon dioxide has an advantage in this respect, as it kills organisms through hypercarbia and hypoxia (depending on concentration), but it requires concentrations of above 35%, or so. This makes carbon dioxide preferable for fumigation in situations where a hermetic seal cannot be maintained.

Controlled Atmospheric Storage (CA): "CA storage is a non-chemical process. Oxygen levels in the sealed rooms are reduced, usually by the infusion of nitrogen gas, from the approximate 21 percent in the air we breathe to 1 percent or 2 percent. Temperatures are kept at a constant 0 to 2 °C (32 to 36 °F). Humidity is maintained at 95 percent and carbon dioxide levels are also controlled. Exact conditions in the rooms are set according to the apple variety. Researchers develop specific regimens for each variety to achieve the best quality. Computers help keep conditions constant." "Eastern Washington, where most of Washington's apples are grown, has enough warehouse storage for 181 million boxes of fruit, according to a report done in 1997 by managers for the Washington State Department of Agriculture Plant Services Division. The storage capacity study shows that 67 percent of that space —enough for 121,008,000 boxes of apples — is CA storage."

Air-tight storage of grains (sometimes called hermetic storage) relies on the respiration of grain, insects, and fungi that can modify the enclosed atmosphere sufficiently to control insect pests. This is a method of great antiquity, as well as having modern equivalents. The success of the method relies on having the correct mix of sealing, grain moisture, and temperature.

A patented process uses fuel cells to exhaust and automatically maintain the exhaustion of oxygen in a shipping container, containing, for example, fresh fish.

Nonthermal Plasma

This process subjects the surface of food to a "flame" of ionized gas molecules, such as helium or nitrogen. This causes micro-organisms to die off on the surface.

High-pressure Food Preservation

High-pressure food preservation or pascalization refers to the use of a food preservation technique that makes use of high pressure. "Pressed inside a vessel exerting 70,000 pounds per square inch

(480 MPa) or more, food can be processed so that it retains its fresh appearance, flavor, texture and nutrients while disabling harmful microorganisms and slowing spoilage." By 2005, the process was being used for products ranging from orange juice to guacamole to deli meats and widely sold.

Biopreservation

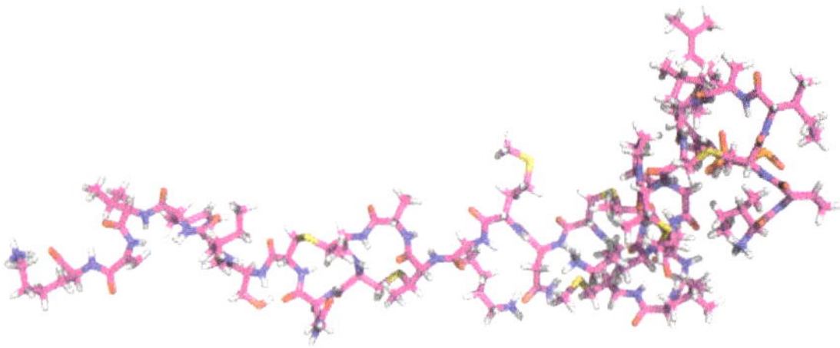

3D stick model of nisin. Some lactic acid bacteria manufacture nisin. It is a particularly effective preservative.

Biopreservation is the use of natural or controlled microbiota or antimicrobials as a way of preserving food and extending its shelf life. Beneficial bacteria or the fermentation products produced by these bacteria are used in biopreservation to control spoilage and render pathogens inactive in food. It is a benign ecological approach which is gaining increasing attention.

Of special interest are lactic acid bacteria (LAB). Lactic acid bacteria have antagonistic properties that make them particularly useful as biopreservatives. When LABs compete for nutrients, their metabolites often include active antimicrobials such as lactic acid, acetic acid, hydrogen peroxide, and peptide bacteriocins. Some LABs produce the antimicrobial nisin, which is a particularly effective preservative.

These days, LAB bacteriocins are used as an integral part of hurdle technology. Using them in combination with other preservative techniques can effectively control spoilage bacteria and other pathogens, and can inhibit the activities of a wide spectrum of organisms, including inherently resistant Gram-negative bacteria.

Hurdle Technology

Hurdle technology is a method of ensuring that pathogens in food products can be eliminated or controlled by combining more than one approach. These approaches can be thought of as "hurdles" the pathogen has to overcome if it is to remain active in the food. The right combination of hurdles can ensure all pathogens are eliminated or rendered harmless in the final product.

Hurdle technology has been defined by Leistner (2000) as an intelligent combination of hurdles that secures the microbial safety and stability as well as the organoleptic and nutritional quality and the economic viability of food products. The organoleptic quality of the food refers to its sensory properties, that is its look, taste, smell, and texture.

Examples of hurdles in a food system are high temperature during processing, low temperature during storage, increasing the acidity, lowering the water activity or redox potential, and the presence of preservatives or biopreservatives. According to the type of pathogens and how risky they are, the intensity of the hurdles can be adjusted individually to meet consumer preferences in an economical way, without sacrificing the safety of the product.

Principal hurdles used for food preservation (after Leistner, 1995)		
Parameter	**Symbol**	**Application**
High temperature	F	Heating
Low temperature	T	Chilling, freezing
Reduced water activity	a_w	Drying, curing, conserving
Increased acidity	pH	Acid addition or formation
Reduced redox potential	E_h	Removal of oxygen or addition of ascorbate
Biopreservatives		Competitive flora such as microbial fermentation
Other preservatives		Sorbates, sulfites, nitrites

References

- Marotz, Lynn R. (2008). Health, Safety, and Nutrition for the Young Child. Wadsworth Publishing. pp. 491–2. ISBN 978-1-4283-2070-3.
- Mitchell, Deborah (2004). Safe foods: the A-to-Z guide to the most wholesome foods for you and your family. Penguin. pp. Ch. 15. ISBN 978-1-101-21015-4.
- Editors of Creative Publishing. Dressing & Cooking Wild Game. Publisher: Creative Publishing international 1999 ISBN 978-0865731080
- Alzamora SM, Tapia MS and López-Malo A (2000) Minimally processed fruits and vegetables: fundamental aspects and applications Springer, Page 266. ISBN 978-0-8342-1672-3.
- Alasalvar C (2010) Seafood Quality, Safety and Health Applications John Wiley and Sons, Page 203. ISBN 978-1-4051-8070-2.
- Leistner L (1995) "Principles and applications of hurdle technology" In Gould GW (Ed.) New Methods of Food Preservation, Springer, pp. 1-21. ISBN 978-0-8342-1341-8.
- "IARC Monographs evaluate consumption of red meat and processed meat" (PDF). International Agency for Research on Cancer. 26 October 2015.
- Laine Welch (May 18, 2013). "Laine Welch: Fuel cell technology boosts long-distance fish shipping". Anchorage Daily News. Retrieved May 19, 2013.
- "Fact Sheet: Freezing and Food Safety". United States Department of Agriculture, Food Safety and Inspection Service. June 3, 2010. Retrieved November 8, 2011.

3

Preservation Methods for Food

Some of the methods of preservation related to food safety are food drying, refrigeration, frozen food, pickling and antioxidants. Food drying is a method in which food is dried, which stops the growth of bacteria and yeasts. Preservation of food by freezing is also a common practice, freezing food preserves it from the time it is prepared, till the time it is eaten. This chapter discusses the methods of food safety in a critical manner providing key analysis to the subject matter.

Food Drying

Food drying is a method of food preservation in which food is dried (dehydrated or desiccated). Drying inhibits the growth of bacteria, yeasts, and mold through the removal of water. Dehydration has been used widely for this purpose since ancient times; the earliest known practice is 12,000 B.C. by inhabitants of the modern Middle East and Asia regions. Water is traditionally removed through evaporation (air drying, sun drying, smoking or wind drying), although today electric food dehydrators or freeze-drying can be used to speed the drying process and ensure more consistent results.

Flattened fish drying in the sun in Madagascar. Fish are preserved through such traditional methods as drying, smoking and salting.

Food Types

Many different foods can be prepared by dehydration. Meat has held an historically significant role. For centuries, much of the European diet depended on dried cod—known as salt cod, bacalhau (with salt), or stockfish (without). It formed the main protein source for the slaves on the West Indian plantations, and was a major economic force within the triangular trade. Dried shark meat, known as Hákarl, is a delicacy in Iceland, while dried reindeer meat is a traditional Sami food. Currently popular dried meats include prosciutto (a.k.a. Parma ham), bresaola, and beef jerky.

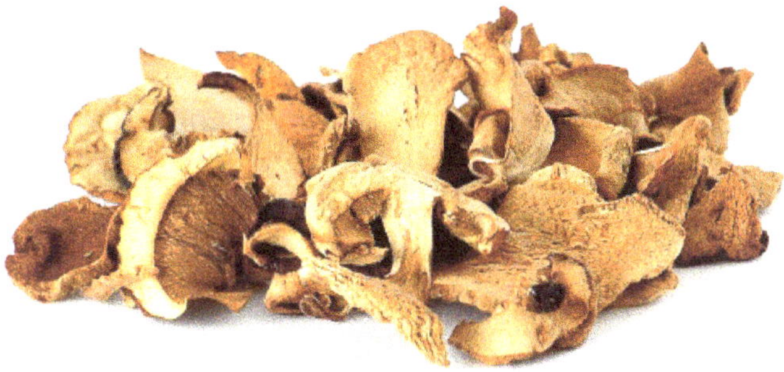

A collection of dried mushrooms

Fruits change character completely when dried. The plum becomes a prune, the grape a raisin. Figs and dates are also transformed into new, different products that can either be eaten as they are or rehydrated. However, dehydrated fruit may be nutritionally inferior to its source.

Sun-drying octopus

Freeze-dried vegetables are often found in food for backpackers, hunters, and the military. Garlic and onion are often dried. Edible mushrooms, as well as other fungi, are also sometimes dried for preservation purposes or to be used as seasonings.

Preparation

Home drying of vegetables, fruit and meat can be carried out with electrical dehydrators (household appliance) or by sun-drying or by wind. Preservatives such as potassium metabisulfite, BHA, or BHT may be used, but are not required. However, dried products without these preservatives may require refrigeration or freezing to ensure safe storage for a long time.

Industrial food dehydration is often accomplished by freeze-drying. In this case food is flash frozen and put into a reduced-pressure system which causes the water to sublimate directly from the solid to the gaseous phase. Although freeze-drying is more expensive than traditional dehydration techniques, it also mitigates the change in flavor, texture, and nutritional value.In addition, another widely used industrial method of drying of food is convective hot air drying. Industrial hot air dryers are simple and easy to design, construct and maintain. More so, it is very affordable and has been reported to retain most of the nutritional properties of food if dry using appropriate drying conditions.

Other Methods

This electric food dehydrator has a hot air blower that blows air through trays with foods on them. Pictured are mango and papaya slices being dried.

There are many different methods for drying, each with their own advantages for particular applications; these include:

- Convection drying
- Bed dryers
- Drum drying
- Freeze Drying
- Microwave-vacuum drying
- Shelf dryers
- Spray drying
- Infrared radiation drying
- Sunlight
- Commercial food dehydrators
- Household oven

Refrigeration

Refrigeration is a process of moving heat from one location to another in controlled conditions. The work of heat transport is traditionally driven by mechanical work, but can also be driven by heat, magnetism, electricity, laser, or other means. Refrigeration has many applications, including, but not limited to: household refrigerators, industrial freezers, cryogenics, and air conditioning. Heat pumps may use the heat output of the refrigeration process, and also may be designed to be reversible, but are otherwise similar to air conditioning units.

Commercial refrigeration

Refrigeration has had a large impact on industry, lifestyle, agriculture and settlement patterns. The idea of preserving food dates back to at least the ancient Roman and Chinese empires. However, mechanical refrigeration technology has rapidly evolved in the last century, from ice harvesting to temperature-controlled rail cars. The introduction of refrigerated rail cars contributed to the westward expansion of the United States, allowing settlement in areas that were not on main transport channels such as rivers, harbors, or valley trails. Settlements were also developing in infertile parts of the country, filled with new natural resources. These new settlement patterns sparked the building of large cities which are able to thrive in areas that were otherwise thought to be inhospitable, such as Houston, Texas and Las Vegas, Nevada. In most developed countries, cities are heavily dependent upon refrigeration in supermarkets, in order to obtain their food for daily consumption. The increase in food sources has led to a larger concentration of agricultural sales coming from a smaller percentage of existing farms. Farms today have a much larger output per person in comparison to the late 1800s. This has resulted in new food sources available to entire populations, which has had a large impact on the nutrition of society.

History

Earliest forms of Cooling

The seasonal harvesting of snow and ice is an ancient practice estimated to have begun earlier than 1000 B.C. A Chinese collection of lyrics from this time period known as the Shih king, describes religious ceremonies for filling and emptying ice cellars. However, little is known about the construction of these ice cellars or what the ice was used for. The next ancient society to harvest ice may have been the Jews according to the book of Proverbs, which reads, "As the cold of snow in the time of harvest, so is a faithful messenger to them who sent him." Historians have interpreted this to mean that the Jews used ice to cool beverages rather than to preserve food. Other ancient cultures such as the Greeks and the Romans dug large snow pits insulated with grass, chaff, or branches of trees as cold storage. Like the Jews, the Greeks and Romans did not use ice and snow to preserve food, but primarily as a means to cool beverages. The Egyptians also developed methods to cool beverages, but in lieu of using ice to cool water, the Egyptians cooled water by putting boiling water in shallow earthen jars and placing them on the roofs of their houses at night. Slaves would moisten the outside of the jars and the resulting evaporation would cool the water. The ancient people of India used this same concept to produce ice. The Persians stored ice in a pit called a Yakhchal and may have been the first group of people to use cold storage to preserve food. In the Australian outback before a reliable electricity supply was available where the weather could be hot and dry, many farmers used a "Coolgardie safe". This consisted of a room with hessian "curtains" hanging from the ceiling soaked in water. The water would evaporate and thereby cool the hessian curtains and thereby the air circulating in the room. This would allow many perishables such as fruit, butter, and cured meats to be kept that would normally spoil in the heat.

Ice Harvesting

Ice harvesting in Massachusetts, 1852, showing the railroad line in the background, used to transport the ice.

Before 1830, few Americans used ice to refrigerate foods due to a lack of ice-storehouses and iceboxes. As these two things became more widely available, individuals used axes and saws to harvest ice for their storehouses. This method proved to be difficult, dangerous, and certainly did not resemble anything that could be duplicated on a commercial scale.

Despite the difficulties of harvesting ice, Frederic Tudor thought that he could capitalize on this new commodity by harvesting ice in New England and shipping it to the Caribbean islands as well as the southern states. In the beginning, Tudor lost thousands of dollars, but eventually turned a profit as he constructed icehouses in Charleston, Virginia and in the Cuban port town of Havana. These icehouses as well as better insulated ships helped reduce ice wastage from 66% to 8%. This efficiency gain influenced Tudor to expand his ice market to other towns with icehouses such as New Orleans and Savannah. This ice market further expanded as harvesting ice became faster and cheaper after one of Tudor's suppliers, Nathaniel Wyeth, invented a horse-drawn ice cutter in 1825. This invention as well as Tudor's success inspired others to get involved in the ice trade and the ice industry grew.

Ice became a mass-market commodity by the early 1830s with the price of ice dropping from six cents per pound to a half of a cent per pound. In New York City, ice consumption increased from 12,000 tons in 1843 to 100,000 tons in 1856. Boston's consumption leapt from 6,000 tons to

85,000 tons during that same period. Ice harvesting created a "cooling culture" as majority of people used ice and iceboxes to store their dairy products, fish, meat, and even fruits and vegetables. These early cold storage practices paved the way for many Americans to accept the refrigeration technology that would soon take over the country.

Refrigeration Research

William Cullen, the first to conduct experiments into artificial refrigeration.

The history of artificial refrigeration began when Scottish professor William Cullen designed a small refrigerating machine in 1755. Cullen used a pump to create a partial vacuum over a container of diethyl ether, which then boiled, absorbing heat from the surrounding air. The experiment even created a small amount of ice, but had no practical application at that time.

In 1758, Benjamin Franklin and John Hadley, professor of chemistry, collaborated on a project investigating the principle of evaporation as a means to rapidly cool an object at Cambridge University, England. They confirmed that the evaporation of highly volatile liquids, such as alcohol and ether, could be used to drive down the temperature of an object past the freezing point of water. They conducted their experiment with the bulb of a mercury thermometer as their object and with a bellows used to "quicken" the evaporation; they lowered the temperature of the thermometer bulb down to 7 °F (−14 °C), while the ambient temperature was 65 °F (18 °C). They noted that soon after they passed the freezing point of water (32 °F), a thin film of ice formed on the surface of the thermometer's bulb and that the ice mass was about a quarter inch thick when they stopped the experiment upon reaching 7 °F (−14 °C). Franklin wrote, "From this experiment, one may see the possibility of freezing a man to death on a warm summer's day". In 1805, American inventor Oliver Evans described a closed vapor-compression refrigeration cycle for the production of ice by ether under vacuum.

In 1820, the English scientist Michael Faraday liquefied ammonia and other gases by using high pressures and low temperatures, and in 1834, an American expatriate to Great Britain, Jacob Perkins, built the first working vapor-compression refrigeration system in the world. It was a closed-cycle that could operate continuously, as he described in his patent:

> I am enabled to use volatile fluids for the purpose of producing the cooling or freezing of fluids, and yet at the same time constantly condensing such volatile fluids, and bringing them again into operation without waste.

His prototype system worked although it did not succeed commercially.

In 1842, a similar attempt was made by American physician, John Gorrie, who built a working prototype, but it was a commercial failure. Like many of the medical experts during this time, Gorrie thought too much exposure to tropical heat led to mental and physical degeneration, as well as the spread of diseases such as malaria. He conceived the idea of using his refrigeration system to cool the air for comfort in homes and hospitals to prevent disease. American engineer Alexander Twining took out a British patent in 1850 for a vapour compression system that used ether.

The first practical vapor compression refrigeration system was built by James Harrison, a British journalist who had emigrated to Australia. His 1856 patent was for a vapour compression system using ether, alcohol or ammonia. He built a mechanical ice-making machine in 1851 on the banks of the Barwon River at Rocky Point in Geelong, Victoria, and his first commercial ice-making machine followed in 1854. Harrison also introduced commercial vapour-compression refrigeration to breweries and meat packing houses, and by 1861, a dozen of his systems were in operation. He later entered the debate of how to compete against the American advantage of unrefrigerated beef sales to the United Kingdom. In 1873 he prepared the sailing ship *Norfolk* for an experimental beef shipment to the United Kingdom, which used a cold room system instead of a refrigeration system. The venture was a failure as the ice was consumed faster than expected.

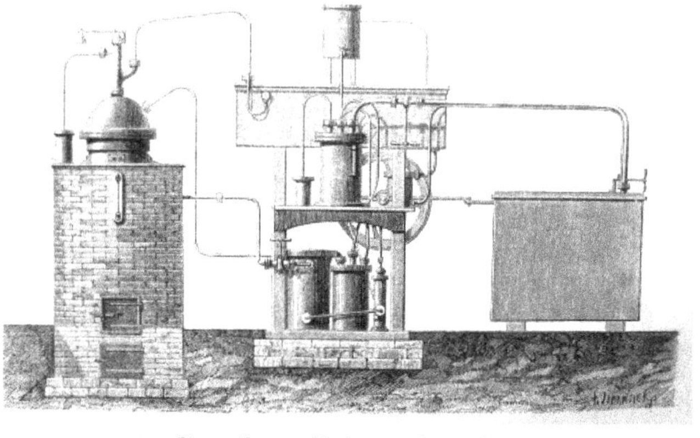

Ferdinand Carré's ice-making device

The first gas absorption refrigeration system using gaseous ammonia dissolved in water (referred to as "aqua ammonia") was developed by Ferdinand Carré of France in 1859 and patented in 1860. Carl von Linde, an engineer specializing in steam locomotives and professor of engineering at the Technological University of Munich in Germany, began researching refrigeration in the 1860s and 1870s in response to demand from brewers for a technology that would allow year-round, large-scale production of lager; he patented an improved method of liquefying gases in 1876. His new process made possible using gases such as ammonia, sulfur dioxide (SO_2) and methyl chloride (CH_3Cl) as refrigerants and they were widely used for that purpose until the late 1920s.

Thaddeus Lowe, an American balloonist, held several patents on ice-making machines. His "Compression Ice Machine" would revolutionize the cold-storage industry. In 1869, other investors and he purchased an old steamship onto which they loaded one of Lowe's refrigeration units and began shipping fresh fruit from New York to the Gulf Coast area, and fresh meat from Galveston, Texas back to New York, but because of Lowe's lack of knowledge about shipping, the business was a costly failure.

Commercial Use

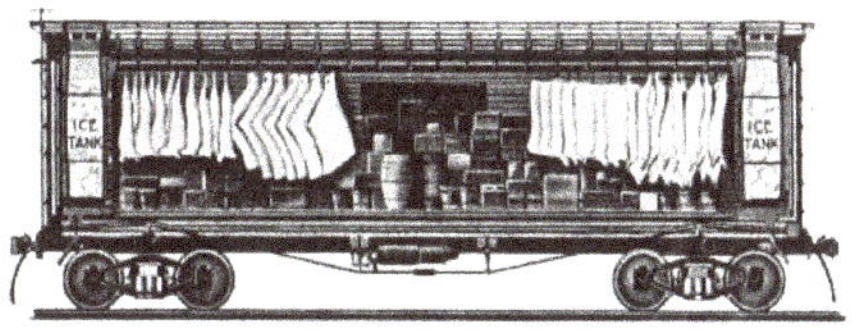

An 1870 refrigerator car design. Hatches in the roof provided access to the tanks for the storage of harvested ice at each end.

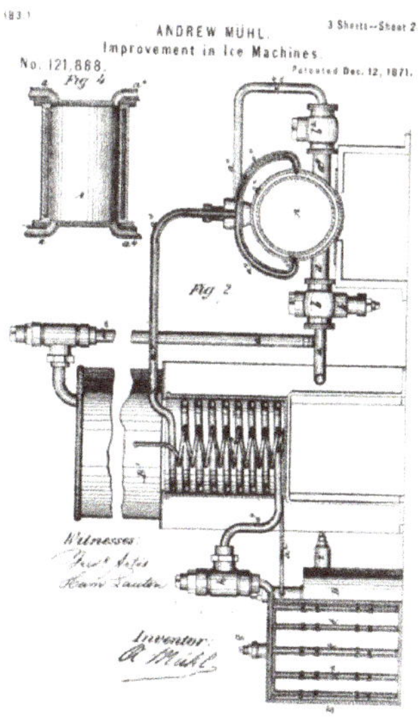

Icemaker Patent by Andrew Muhl, dated December 12, 1871.

In 1842, John Gorrie created a system capable of refrigerating water to produce ice. Although it was a commercial failure, it inspired scientists and inventors around of the world. France's Ferdinand Carre was one of the inspired and he created an ice producing system that was simpler and smaller than that of Gorrie. During the Civil War, cities such as New Orleans could no longer get ice from New England via the coastal ice trade. Carre's refrigeration system became the solution to New Orleans ice problems and by 1865 the city had three of Carre's machines. In 1867, in San Antonio, Texas, a French immigrant named Andrew Muhl built an ice-making machine to help service the expanding beef industry before moving it to Waco in 1871. In 1873, the patent for this machine was contracted by the Columbus Iron Works, a company acquired by the W. C. Bradley Co., which went on to produce the first commercial ice-makers in the US.

By the 1870s, breweries had become the largest users of harvested ice. Though the ice-harvesting industry had grown immensely by the turn of the 20th century, pollution and sewage had begun to creep into natural ice, making it a problem in the metropolitan suburbs. Eventually, breweries began to complain of tainted ice. Public concern for the purity of water, from which ice was formed, began to increase in the early 1900s with the rise of germ theory. Numerous media outlets published articles connecting diseases such as typhoid fever with natural ice consumption. This caused ice harvesting to become illegal in certain areas of the country. All of these scenarios increased the demands for modern refrigeration and manufactured ice. Ice producing machines like that of Carre's and Muhl's were looked to as means of producing ice to meet the needs of grocers, farmers, and food shippers.

Refrigerated railroad cars were introduced in the US in the 1840s for short-run transport of dairy products, but these used harvested ice to maintain a cool temperature.

Dunedin, the first commercially successful refrigerated ship.

The new refrigerating technology first met with widespread industrial use as a means to freeze meat supplies for transport by sea from the British Dominions and other countries to the British Isles. The first to achieve this breakthrough was an entrepreneur who had emigrated to New Zealand. William Soltau Davidson thought that Britain's rising population and meat demand could mitigate the slump in world wool markets that was heavily affecting New Zealand. After extensive research, he commissioned the *Dunedin* to be refitted with a compression refrigeration unit for meat shipment in 1881. On February 15, 1882, the *Dunedin* sailed for London with what was to be the first commercially successful refrigerated shipping voyage, and the foundation of the refrigerated meat industry.

The Times commented "Today we have to record such a triumph over physical difficulties, as would have been incredible, even unimaginable, a very few days ago...". The *Marlborough*—sister ship to the *Dunedin* – was immediately converted and joined the trade the following year, along with the rival New Zealand Shipping Company vessel *Mataurua*, while the German Steamer *Marsala* began carrying frozen New Zealand lamb in December 1882. Within five years, 172 shipments of frozen meat were sent from New Zealand to the United Kingdom, of which only 9 had significant amounts of meat condemned. Refrigerated shipping also led to a broader meat and dairy boom in Australasia and South America. J & E Hall of Dartford, England outfitted the 'SS Selembria' with a vapor compression system to bring 30,000 carcasses of mutton from the Falkland Islands in 1886. In the years ahead, the industry rapidly expanded to Australia, Argentina and the United States.

By the 1890s, refrigeration played a vital role in the distribution of food. The meat-packing industry relied heavily on natural ice in the 1880s and continued to rely on manufactured ice as those technologies became available. By 1900, the meat-packing houses of Chicago had adopted ammonia-cycle commercial refrigeration. By 1914, almost every location used artificial refrigeration. The big meat-packers, Armour, Swift, and Wilson, had purchased the most expensive units which they installed on train cars and in branch houses and storage facilities in the more remote distribution areas.

By the middle of the 20th century, refrigeration units were designed for installation on trucks or lorries. Refrigerated vehicles are used to transport perishable goods, such as frozen foods, fruit and vegetables, and temperature-sensitive chemicals. Most modern refrigerators keep the temperature between −40 and −20 °C, and have a maximum payload of around 24,000 kg gross weight (in Europe).

Although commercial refrigeration quickly progressed, it had limitations that prevented it from moving into the household. First, most refrigerators were far too large. Some of the commercial units being used in 1910 weighed between five and two hundred tons. Second, commercial refrigerators were expensive to produce, purchase, and maintain. Lastly, these refrigerators were unsafe. It was not uncommon for commercial refrigerators to catch fire, explode, or leak toxic gases. Refrigeration did not become a household technology until these three challenges were overcome.

Home and Consumer Use

During the early 1800s, consumers preserved their food by storing food and ice purchased from ice harvesters in iceboxes. In 1803, Thomas Moore patented a metal-lined butter-storage tub which became the prototype for most iceboxes. These iceboxes were used until nearly 1910 and the technology did not progress. In fact, consumers that used the icebox in 1910 faced the same challenge of a moldy and stinky icebox that consumers had in the early 1800s.

General Electric (GE) was one of the first companies to overcome these challenges. In 1911, GE released a household refrigeration unit that was powered by gas. The use of gas eliminated the need for motor and decreased the size of the refrigerator. However, electric companies that were customers of GE did not benefit from a gas-powered unit. Thus, GE invested in developing an electric model. In 1927, GE released the Monitor Top, the first refrigerator to run off electricity.

In 1930, Frigidaire, one of GE's main competitors, synthesized Freon. With the invention of synthetic refrigerants based mostly on a chlorofluorocarbon (CFC) chemical, safer refrigerators were possible for home and consumer use. Freon led to the development of smaller, lighter, and cheaper refrigerators. The average price of a refrigerator dropped from $275 to $154 with the synthesis of Freon. This lower price allowed ownership of refrigerators in American households to exceed 50%. Freon is a trademark of the DuPont Corporation and refers to these CFCs, and later hydro chlorofluorocarbon (HCFC) and hydro fluorocarbon (HFC), refrigerants developed in the late 1920s. These refrigerants were considered at the time to be less harmful than the commonly-used refrigerants of the time, including methyl formate, ammonia, methyl chloride, and sulfur dioxide. The intent was to provide refrigeration equipment for home use without danger. These CFC refrigerants answered that need. In the 1970s, though, the compounds were found to be reacting with atmospheric ozone, an important protection against solar ultraviolet radiation, and their use as a refrigerant worldwide was curtailed in the Montreal Protocol of 1987.

Impact on Settlement Patterns

In the last century refrigeration allowed new settlement patterns to emerge. This new technology has allowed for new areas to be settled that are not on a natural channel of transport such as a river, valley trail or harbor that may have otherwise not been settled. Refrigeration has given opportunities to early settlers to expand westward and into rural areas that were unpopulated. These new settlers with rich and untapped soil saw opportunity to profit by sending raw goods to the eastern cities and states. In the 20th century, refrigeration has made "Galactic Cities" such as Dallas, Phoenix and Los Angeles possible.

Refrigerated Rail Cars

The refrigerated rail car, along with the dense railroad network, became an exceedingly important link between the marketplace and the farm allowing for a national opportunity rather than a just a regional one. Before the invention of the refrigerated rail car it was impossible to ship perishable food products long distances. The beef packing industry made the first demand push for refrigeration cars. The railroad companies were slow to adopt this new invention because of their heavy investments in cattle cars, stockyards, and feedlots. Refrigeration cars were also complex and costly compared to other rail cars, which also slowed the adoption of the refrigerated rail car. After the slow adoption of the refrigerated car, the beef packing industry dominated the refrigerated rail car business with their ability to control ice plants and the setting of icing fees. The United States Department of Agriculture estimated that in 1916 over sixty-nine percent of the cattle killed in the country was done in plants involved in interstate trade. The same companies that were also involved in the meat trade later implemented refrigerated transport to include vegetables and fruit. The meat packing companies had much of the expensive machinery, such as refrigerated cars, and cold storage facilities that allowed for them to effectively distribute all types of perishable goods. During World War I, a national refrigerator car pool was established by the United States Administration to deal with problem of idle cars and was later continued after the war. The idle car problem was the problem of refrigeration cars sitting pointlessly in between seasonal harvests. This meant that very expensive cars sat in rail yards for a good portion of the year while making no revenue for the car's owner. The car pool was a system where cars were distributed to areas as crops matured ensuring maximum use of the cars. Refrigerated rail cars moved eastward from vineyards, orchards, fields, and gardens in western states to satisfy Americas consuming market in the east. The refrigerated car made it possible to transport perishable crops hundreds and even thousands of miles. The most noticeable effect the car gave was a regional specialization of vegetables and fruits. The refrigeration rail car was widely used for the transportation of perishable goods up until the 1950s. By the 1960s the nation's interstate highway system was adequately complete allowing for trucks to carry the majority of the perishable food loads and to push out the old system of the refrigerated rail cars.

Expansion West and into Rural Areas

The widespread use of refrigeration allowed for a vast amount of new agricultural opportunities to open up in the United States. New markets emerged throughout the United States in areas that were previously uninhabited and far-removed from heavily populated areas. New agricultural opportunity presented itself in areas that were considered rural such as states in the south and in the west. Shipments on a large scale from the south and California were both made around the same time although natural ice was used from the Sierras in California rather than manufactured ice in the south. Refrigeration allowed for many areas to specialize in the growing of specific fruits. California specialized in several fruits, grapes, peaches, pears, plums, and apples while Georgia became famous for specifically its peaches. In California, the acceptance of the refrigerated rail carts lead to an increase of car loads from 4,500 carloads in 1895 to between 8,000 and 10,000 carloads in 1905. The Gulf States, Arkansas, Missouri and Tennessee entered into strawberry production on a large-scale while Mississippi became the center of the tomato industry. New Mexico, Colorado, Arizona, and Nevada grew cantaloupes. Without refrigeration this would have not been possible. By 1917, well-established fruit and vegetable areas that were close to eastern markets felt the pressure of competition from these distant specialized centers. Refrigeration was not limited to meat, fruit and vegetables but it also encompassed dairy product and dairy farms. In the early twentieth century large cities got their dairy supply from farms as far as 400 miles. Dairy products were not as easily transported great distances like fruits and vegetables due to greater perishability. Refrigeration made production possible in the west far from eastern markets, so much in fact that dairy farmers could pay transportation cost and still undersell their eastern competitors. Refrigeration and the refrigerated rail gave opportunity to areas with rich soil far from natural channel of transport such as a river, valley trail or harbors.

Rise of the Galactic City

"Edge city" was a term coined by Joel Garreau, whereas the term "galactic city" was coined by Lewis Mumford. These terms refer to a concentration of business, shopping, and entertainment outside a traditional downtown or central business district in what had previously been a residential or rural area. There were several factors contributing to the growth of these cities such as Los Angeles, Las Vegas, Houston, and Phoenix. The factors that contributed to these large cities include reliable automobiles, highway systems, refrigeration, and agricultural production increases. Large cities such as the ones mentioned above have not been uncommon in history but what separates these cities from the rest are that these cities are not along some natural channel of transport, or at some crossroad of two or more channels such as a trail, harbor, mountain, river, or valley. These large cities have been developed in areas that only a few hundred years ago would have been uninhabitable. Without a cost efficient way of cooling air and transporting water and food great distances these large cities would have never developed. The rapid growth of these cities was influenced by refrigeration and an agricultural productivity increase, allowing more distant farms to effectively feed the population.

Impact on Agriculture and Food Production

Agriculture's role in developed countries has drastically changed in the last century due to many factors, including refrigeration. Statistics from the 2007 census gives information on the large concentration of agricultural sales coming from a small portion of the existing farms in the United States today. This is a partial result of the market created for the frozen meat trade by the first successful shipment of frozen sheep carcasses coming from New Zealand in the 1880s. As the market continued to grow, regulations on food processing and quality began to be enforced. Eventually, electricity was introduced into rural homes in the United States, which allowed refrigeration technology to continue to expand on the farm, increasing output per person. Today, refrigeration's use on the farm reduces humidity levels, avoids spoiling due to bacterial growth, and assists in preservation.

Demographics

The introduction of refrigeration and evolution of additional technologies drastically changed agriculture in the United States. During the beginning of the 20th century, farming was a common occupation and lifestyle for United States citizens, as most farmers actually lived on their farm. In 1935, there were 6.8 million farms in the United States and a population of 127 million. Yet, while the United States population has continued to climb, citizens pursuing agriculture continue to decline. Based on the 2007 US Census, less than one percent of a population of 310 million people claim farming as an occupation today. However, the increasing population has led to an increasing demand for agricultural products, which is met through a greater variety of crops, fertilizers, pesticides, and improved technology. Improved technology has decreased the risk and time involved if agricultural management and allows larger farms to increase their output per person to meet society's demand.

Meat Packing and Trade

Prior to 1882, the South Island of New Zealand had been experimenting with sowing grass and crossbreeding sheep, which immediately gave their farmers economic potential in the exportation of meat. In 1882, the first successful shipment of sheep carcasses was sent from Port Chalmers in Dunedin, New Zealand to London. By the 1890s, the frozen meat trade became increasingly more profitable in New Zealand, especially in Canterbury, where 50% of exported sheep carcasses came from in 1900. It wasn't long before Canterbury meat was known for the highest quality, creating a demand for New Zealand meat around the world. In order to meet this new demand, the farmers improved their feed so sheep to be ready for the slaughter in only seven months. This new method of shipping lead to an economic boom in New Zealand by the mid 1890s.

In the United States, the Meat Inspection Act of 1891 was put in place in the United States because local butchers felt the refrigerated railcar system was unwholesome. When meat packing began to take off, consumers became nervous about the quality of the meat for consumption. The Jungle, a book written by Upton Sinclair, brought negative attention to the meat packing industry, by drawing to light unsanitary working conditions and processing of diseased animals. The book caught the attention of President Theodore Roosevelt, and the 1906 Meat Inspection Act was put into place as an amendment to the Meat Inspection Act of 1891. This new act focused on the quality of the meat and environment it is processed in.

Electricity in Rural Areas

In the early 1930s, 90 percent of the urban population had electric power, in comparison to only

10 percent of rural homes. At the time, power companies did not feel that extending power to rural areas would produce enough profit to make it worth their while. However, in the midst of the Great Depression, President Franklin D. Roosevelt realized that rural areas would continue to lag behind urban areas in both poverty and production if they were not electrically wired. On May 11, 1935, the president signed an executive order called the Rural Electrification Administration, also known as REA. The agency provided loans to fund electric infrastructure in the rural areas. In just a few years, 300,000 people in rural areas of the United States had received power in their homes.

While electricity dramatically improved working conditions on farms, it also had a large impact on the safety of food production. Refrigeration systems were introduced to the farming process, which helped in food preservation and kept food supplies safe. Refrigeration also allowed for production of perishable commodities, which could then be shipped throughout the United States. As a result, the United States farmers quickly became the most productive in the world.

Farm Use

In order to reduce humidity levels and spoiling due to bacterial growth, refrigeration is used for meat, produce, and dairy processing in farming today. Refrigeration systems are used the heaviest in the warmer months for farming produce, which must be cooled as soon as possible in order to meet quality standards and increase the shelf life. Meanwhile, dairy farms refrigerate milk year round to avoid spoiling.

Effects on Lifestyle and Diet

A trip to the market before refrigeration became widespread would have been different from a trip today. In the late 19th Century and into the very early 20th Century, other than staple foods (sugar, rice, and beans), your diet was affected heavily by the seasons and what could be grown relatively close to your region. Today, thanks to refrigeration, we are no longer restricted by these limitations. Refrigeration played a large part in the feasibility and then popularity of the modern supermarket. If you are willing to pay slightly more for a fruit or vegetable that is out of season in your region, your local supermarket will most likely have what you are looking for. Refrigerators have led to a huge increase in meat and dairy as a portion of overall supermarket sales. As well as changing the goods purchased at the market, the ability to store these foods for extended periods of time has led to an increase in leisure time. Prior to the advent of the household refrigerator, people would have to shop on a daily basis for the supplies needed for their meals.

Impact on Nutrition

The introduction of refrigeration allowed for the hygienic handling and storage of perishables, and as such, promoted output growth, consumption, and nutrition. The change in our method of food preservation moved us away from salts to a more manageable sodium level. The ability to move and store perishables such as meat and dairy led to a 1.7% increase in dairy consumption and overall protein intake by 1.25% annually in the US after the 1890s. People were not only consuming these perishables because it became easier for they themselves to store them, but because the innovations in refrigerated transportation and storage led to less spoilage and waste, and drove the prices of these products down. Refrigeration accounts for at least 5.1% of the increase in adult stature (in the US) through improved nutrition including the indirect effects associated with improvements in the quality of nutrients and the reduction in illness, the overall impact was considerably larger. Recent studies have also shown a negative relationship between the number of refrigerators in a household and the rate of gastric cancer mortality.

Current Applications of Refrigeration

Probably the most widely used current applications of refrigeration are for air conditioning of private homes and public buildings, and refrigerating foodstuffs in homes, restaurants and large storage warehouses. The use of refrigerators in kitchens for storing fruits and vegetables has allowed adding fresh salads to the modern diet year round, and storing fish and meats safely for long periods. Optimum temperature range for perishable food storage is 3 to 5 °C (37 to 41 °F).

In commerce and manufacturing, there are many uses for refrigeration. Refrigeration is used to liquify gases - oxygen, nitrogen, propane and methane, for example. In compressed air purification, it is used to condense water vapor from compressed air to reduce its moisture content. In oil refineries, chemical plants, and petrochemical plants, refrigeration is used to maintain certain processes at their needed low temperatures (for example, in alkylation of butenes and butane to produce a high octane gasoline component). Metal workers use refrigeration to temper steel and cutlery. In transporting temperature-sensitive foodstuffs and other materials by trucks, trains, airplanes and seagoing vessels, refrigeration is a necessity.

Dairy products are constantly in need of refrigeration, and it was only discovered in the past few decades that eggs needed to be refrigerated during shipment rather than waiting to be refrigerated after arrival at the grocery store. Meats, poultry and fish all must be kept in climate-controlled environments before being sold. Refrigeration also helps keep fruits and vegetables edible longer.

One of the most influential uses of refrigeration was in the development of the sushi/sashimi industry in Japan. Before the discovery of refrigeration, many sushi connoisseurs were at risk of contracting diseases. The dangers of unrefrigerated sashimi were not brought to light for decades due to the lack of research and healthcare distribution across rural Japan. Around mid-century, the Zojirushi corporation, based in Kyoto, made breakthroughs in refrigerator designs, making refrigerators cheaper and more accessible for restaurant proprietors and the general public.

Methods of Refrigeration

Methods of refrigeration can be classified as *non-cyclic*, *cyclic*, *thermoelectric* and *magnetic*.

Non-cyclic Refrigeration

This refrigeration method cools a contained area by melting ice, or by sublimating dry ice. Perhaps the simplest example of this is a portable cooler, where items are put in it, then ice is poured over the top. Regular ice can maintain temperatures near, but not below the freezing point, unless salt is used to cool the ice down further (as in a traditional ice-cream maker). Dry ice can reliably bring the temperature well below freezing.

Cyclic Refrigeration

This consists of a refrigeration cycle, where heat is removed from a low-temperature space or source and rejected to a high-temperature sink with the help of external work, and its inverse, the thermodynamic power cycle. In the power cycle, heat is supplied from a high-temperature source to the engine, part of the heat being used to produce work and the rest being rejected to a low-temperature sink. This satisfies the second law of thermodynamics.

A *refrigeration cycle* describes the changes that take place in the refrigerant as it alternately absorbs and rejects heat as it circulates through a refrigerator. It is also applied to heating, ventilation, and air conditioning HVACR work, when describing the "process" of refrigerant flow through an HVACR unit, whether it is a packaged or split system.

Heat naturally flows from hot to cold. Work is applied to cool a living space or storage volume by pumping heat from a lower temperature heat source into a higher temperature heat sink. Insulation is used to reduce the work and energy needed to achieve and maintain a lower temperature in the cooled space. The operating principle of the refrigeration cycle was described mathematically by Sadi Carnot in 1824 as a heat engine.

The most common types of refrigeration systems use the reverse-Rankine vapor-compression refrigeration cycle, although absorption heat pumps are used in a minority of applications.

Cyclic refrigeration can be classified as:

1. Vapor cycle, and
2. Gas cycle

Vapor cycle refrigeration can further be classified as:

1. Vapor-compression refrigeration
2. Vapor-absorption refrigeration

Vapor-compression Cycle

The vapor-compression cycle is used in most household refrigerators as well as in many large commercial and industrial refrigeration systems. Figure 1 provides a schematic diagram of the components of a typical vapor-compression refrigeration system.

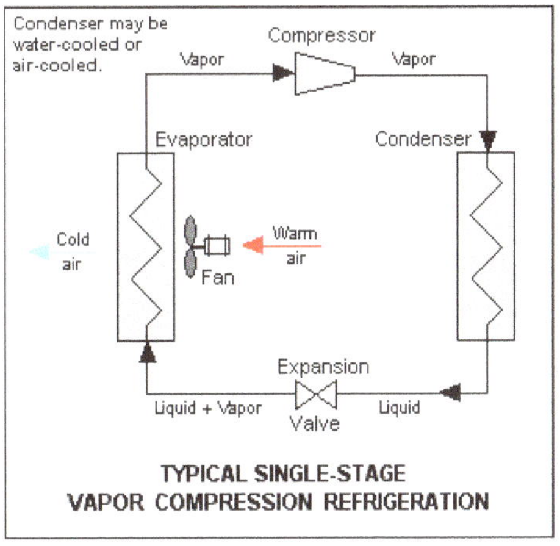

Figure 1: Vapor compression refrigeration

The thermodynamics of the cycle can be analyzed on a diagram as shown in Figure 2. In this cycle, a circulating refrigerant such as Freon enters the compressor as a vapor. From point 1 to point 2, the vapor is compressed at constant entropy and exits the compressor as a vapor at a higher temperature, but still below the vapor pressure at that temperature. From point 2 to point 3 and on to point 4, the vapor travels through the condenser which cools the vapor until it starts condensing, and then condenses the vapor into a liquid by removing additional heat at constant pressure and temperature. Between points 4 and 5, the liquid refrigerant goes through the expansion valve (also called a throttle valve) where its pressure abruptly decreases, causing flash evaporation and auto-refrigeration of, typically, less than half of the liquid.

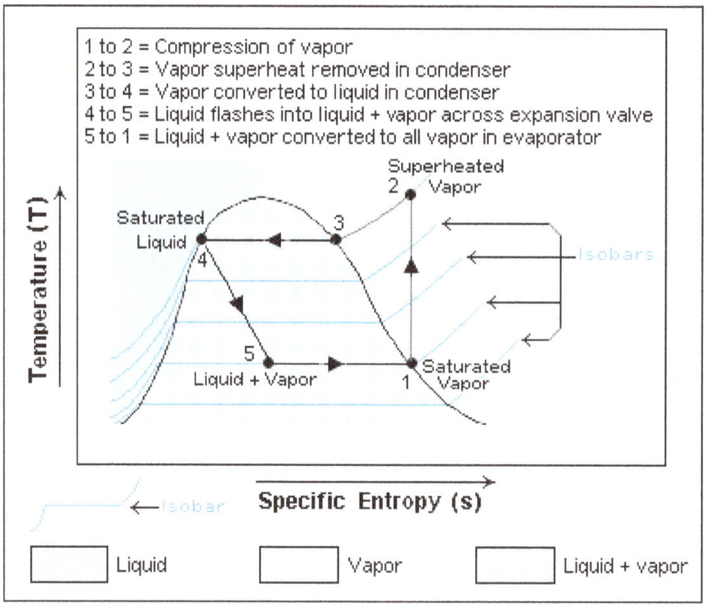

Figure 2: Temperature–Entropy diagram

That results in a mixture of liquid and vapor at a lower temperature and pressure as shown at point 5. The cold liquid-vapor mixture then travels through the evaporator coil or tubes and is completely vaporized by cooling the warm air (from the space being refrigerated) being blown by a fan across the evaporator coil or tubes. The resulting refrigerant vapor returns to the compressor inlet at point 1 to complete the thermodynamic cycle.

The above discussion is based on the ideal vapor-compression refrigeration cycle, and does not take into account real-world effects like frictional pressure drop in the system, slight thermodynamic irreversibility during the compression of the refrigerant vapor, or non-ideal gas behavior, if any.

More information about the design and performance of vapor-compression refrigeration systems is available in the classic *Perry's Chemical Engineers' Handbook*.

Vapor Absorption Cycle

In the early years of the twentieth century, the vapor absorption cycle using water-ammonia systems was popular and widely used. After the development of the vapor compression cycle, the vapor absorption cycle lost much of its importance because of its low coefficient of performance (about one fifth of that of the vapor compression cycle). Today, the vapor absorption cycle is used mainly where fuel for heating is available but electricity is not, such as in recreational vehicles that carry LP gas. It is also used in industrial environments where plentiful waste heat overcomes its inefficiency.

The absorption cycle is similar to the compression cycle, except for the method of raising the pressure of the refrigerant vapor. In the absorption system, the compressor is replaced by an absorber which dissolves the refrigerant in a suitable liquid, a liquid pump which raises the pressure and a generator which, on heat addition, drives off the refrigerant vapor from the high-pressure liquid. Some work is needed by the liquid pump but, for a given quantity of refrigerant, it is much smaller than needed by the compressor in the vapor compression cycle. In an absorption refrigerator, a suitable combination of refrigerant and absorbent is used. The most common combinations are ammonia (refrigerant) with water (absorbent), and water (refrigerant) with lithium bromide (absorbent).

Gas Cycle

When the working fluid is a gas that is compressed and expanded but doesn't change phase, the refrigeration cycle is called a *gas cycle*. Air is most often this working fluid. As there is no condensation and evaporation intended in a gas cycle, components corresponding to the condenser and evaporator in a vapor compression cycle are the hot and cold gas-to-gas heat exchangers in gas cycles.

The gas cycle is less efficient than the vapor compression cycle because the gas cycle works on the reverse Brayton cycle instead of the reverse Rankine cycle. As such the working fluid does not receive and reject heat at constant temperature. In the gas cycle, the refrigeration effect is equal to the product of the specific heat of the gas and the rise in temperature of the gas in the low temperature side. Therefore, for the same cooling load, a gas refrigeration cycle needs a large mass flow rate and is bulky.

Because of their lower efficiency and larger bulk, *air cycle* coolers are not often used nowadays in terrestrial cooling devices. However, the air cycle machine is very common on gas turbine-powered jet aircraft as cooling and ventilation units, because compressed air is readily available from the engines' compressor sections. Such units also serve the purpose of pressurizing the aircraft.

Thermoelectric Refrigeration

Thermoelectric cooling uses the Peltier effect to create a heat flux between the junction of two types of material. This effect is commonly used in camping and portable coolers and for cooling electronic components and small instruments.

Magnetic Refrigeration

Magnetic refrigeration, or adiabatic demagnetization, is a cooling technology based on the magnetocaloric effect, an intrinsic property of magnetic solids. The refrigerant is often a paramagnetic salt, such as cerium magnesium nitrate. The active magnetic dipoles in this case are those of the electron shells of the paramagnetic atoms.

A strong magnetic field is applied to the refrigerant, forcing its various magnetic dipoles to align and putting these degrees of freedom of the refrigerant into a state of lowered entropy. A heat sink then absorbs the heat released by the refrigerant due to its loss of entropy. Thermal contact with the heat sink is then broken so that the system is insulated, and the magnetic field is switched off. This increases the heat capacity of the refrigerant, thus decreasing its temperature below the temperature of the heat sink.

Because few materials exhibit the needed properties at room temperature, applications have so far been limited to cryogenics and research.

Other Methods

Other methods of refrigeration include the air cycle machine used in aircraft; the vortex tube used for spot cooling, when compressed air is available; and thermoacoustic refrigeration using sound waves in a pressurized gas to drive heat transfer and heat exchange; steam jet cooling popular in the early 1930s for air conditioning large buildings; thermoelastic cooling using a smart metal alloy stretching and relaxing. Many Stirling cycle heat engines can be run backwards to act as a refrigerator, and therefore these engines have a niche use in cryogenics. In addition there are other types of cryocoolers such as Gifford-McMahon coolers, Joule-Thomson coolers, pulse-tube refrigerators and, for temperatures between 2 mK and 500 mK, dilution refrigerators.

Fridge Gate

The Fridge Gate method is a theoretical application of using a single logic gate to drive a refrigerator in the most energy efficient way possible without violating the laws of thermodynamics. It operates on the fact that there are two energy states in which a particle can exist: the ground state and the excited state. The excited state carries a little more energy than the ground state, small enough so that the transition occurs with high probability. There are three components or particle types associated with the fridge gate. The first is on the interior of the fridge, the second on the outside and the third is connected to a power supply which heats up every so often that it can reach the E state and replenish the source. In the cooling step on the inside of the fridge, the g state particle absorbs energy from ambient particles, cooling them, and itself jumping to the e state. In the second step, on the outside of the fridge where the particles are also at an e state, the particle falls to the g state, releasing energy and heating the outside particles. In the third and final step, the power supply moves a particle at the e state, and when it falls to the g state it induces an energy-neutral swap where the interior e particle is replaced by a new g particle, restarting the cycle.

Capacity Ratings

The measured capacity of refrigeration is always dimensioned in units of power. Domestic and commercial refrigerators may be rated in kJ/s, or Btu/h of cooling. For commercial and industrial refrigeration systems, most of the world uses the kilowatt (kW) as the basic unit of refrigeration. Typically, commercial and industrial refrigeration systems in North America are rated in tons of refrigeration (TR). Historically, one TR was defined as the energy removal rate that will freeze one short ton of water at 0 °C (32 °F) in one day. This was very important because many early refrigeration systems were in ice houses. The simple unit allowed owners of these early refrigeration systems to measure a day's output of ice against energy consumption, and to compare their plant to one down the street. While ice houses make up a much smaller part of the refrigeration industry than they once did, the unit TR has remained in North America. The unit's value as historically defined was approximately 11,958 Btu/hr (3.505 kW), and has now been conventionally redefined as exactly 12,000 Btu/hr (3.517 kW).

A refrigeration system's coefficient of performance (CoP) is very important in determining a system's overall efficiency. It is defined as refrigeration capacity in kW divided by the energy input in kW. While CoP is a very simple measure of performance, it is typically not used for industrial refrigeration in North America. Owners and manufacturers of these systems typically use performance factor (PF). A system's PF is defined as a system's energy input in horsepower divided by its refrigeration capacity in TR. Both CoP and PF can be applied to either the entire system or to system components. For example, an individual compressor can be rated by comparing the energy needed to run the compressor versus the expected refrigeration capacity based on inlet volume flow rate. It is important to note that both CoP and PF for a refrigeration system are only defined at specific operating conditions, including temperatures and thermal loads. Moving away from the specified operating conditions can dramatically change a system's performance.

Frozen Food

A frozen processed foods aisle at a supermarket in Canada

Freezing food preserves it from the time it is prepared to the time it is eaten. Since early times, farmers, fishermen, and trappers have preserved their grains and produce in unheated buildings during the winter season. Freezing food slows down decomposition by turning residual moisture into ice, inhibiting the growth of most bacterial species. In the food commodity industry, there are two processes: mechanical and cryogenic (or flash freezing). The freezing kinetics is important to preserve the food quality and texture. Quicker freezing generates smaller ice crystals and maintains cellular structure. Cryogenic freezing is the quickest freezing technology available due to the ultra low liquid nitrogen temperature −196 °C (−320 °F).

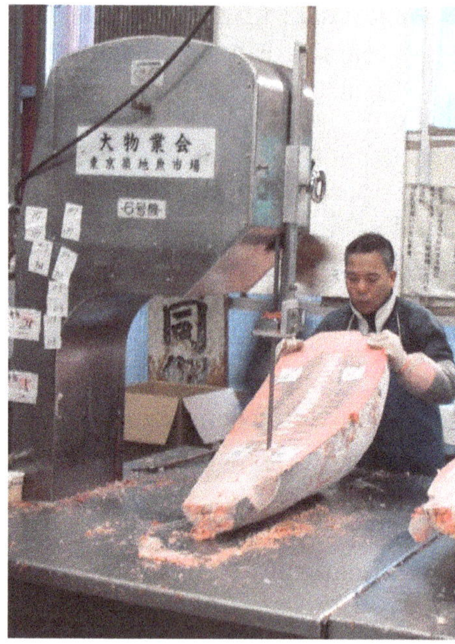

Cutting frozen tuna using a band saw in the Tsukiji fish market in Tokyo, Japan (2002)

Preserving food in domestic kitchens during the 20th and 21st centuries is achieved using household freezers. Accepted advice to householders was to freeze food on the day of purchase. An initiative by a supermarket group in 2012 (backed by the UK's Waste & Resources Action Programme) promotes advising the freezing of food "as soon as possible up to the product's 'use by' date". The Food Standards Agency was reported as supporting the change, providing the food had been stored correctly up to that time.

Preservatives

Frozen products do not require any added preservatives because microorganisms do not grow when the temperature of the food is below −9.5 °C (15 °F), which is sufficient on its own in preventing food spoilage. Long-term preservation of food may call for food storage at even lower temperatures. Carboxymethylcellulose (CMC), a tasteless and odorless stabilizer, is typically added to frozen food because it does not adulterate the quality of the product.

History

Natural food freezing (using winter frosts) had been in use by tribes in cold climates for centuries. By 1885 a small number of chicken and geese were being shipped from Russia to London in insulated cases using this technique. By March 1899, the "British Refrigeration and Allied Interests" reported that a food importing business, "Baerselman Bros", was shipping some 200,000 frozen geese and chickens per week from three Russian depots to New Star Wharf, Lower Shadwell, London over three or four winter months. This trade in frozen food was enabled by the introduction of Linde cold air freezing plants in three Russian depots and the London warehouse. The Shadwell warehouse stored the frozen goods until they were shipped to markets in London, Birmingham, Liverpool and Manchester. The techniques were later expanded into the meat packing industry.

From 1929, Clarence Birdseye introduced "flash freezing" to the American public. Birdseye first became interested in food freezing during fur-trapping expeditions to Labrador in 1912 and 1916, where he saw the natives use natural freezing to preserve foods. More advanced attempts include food frozen for Eleanor Roosevelt on her trip to Russia. Other experiments, involving orange juice, ice cream and vegetables were conducted by the military near the end of World War II.

Technology

The freezing technique itself, just like the frozen food market, is developing to become faster, more efficient and more cost-effective.

Mechanical freezers were the first to be used in the food industry and are used in the vast majority of freezing / refrigerating lines. They function by circulating a refrigerant, normally ammonia, around the system, which withdraws heat from the food product. This heat is then transferred to a condenser and dissipated into air or water. The refrigerant itself, now a high pressure, hot liquid, is directed into an evaporator. As it passes through an expansion valve, it is cooled and then vaporises into a gaseous state. Now a low pressure, low temperature gas again, it can be reintroduced into the system.

Cryogenic or (flash freezing) of food is a more recent development, but is used by many leading food manufacturers all over the world. Cryogenic equipment uses very low temperature gases – usually liquid nitrogen or solid carbon dioxide – which are applied directly to the food product.

Packaging

Frozen food packaging must maintain its integrity throughout machine filling, sealing, freezing, storage, transportation, thawing, and often cooking. As many frozen foods are cooked in a microwave oven, manufacturers have developed packaging that can go straight from freezer to the microwave.

In 1974, the first differential heating container (DHC) was sold to the public. A DHC is a sleeve of metal designed to allow frozen foods to receive the correct amount of heat. Various sized apertures were positioned around the sleeve. The consumer would put the frozen dinner into the sleeve according to what needed the most heat. This ensured proper cooking.

Today there are multiple options for packaging frozen foods. Boxes, cartons, bags, pouches, Boil-in-Bags, lidded trays and pans, crystallized PET trays, and composite and plastic cans.

Scientists are continually researching new aspects of frozen food packaging. Active packaging offers a host of new technologies that can actively sense and then neutralize the presence of bacteria or other harmful species. Active packaging can extend shelf-life, maintain product safety, and help preserve the food over a longer period of time. Several functions of active packaging are being researched:

- Oxygen scavengers
- Time Temperature Indicators and digital temperature data loggers
- Antimicrobials
- Carbon Dioxide controllers

- Microwave susceptors
- Moisture control: Water activity, Moisture vapor transmission rate, etc.
- Flavor enhancers
- Odor generators
- Oxygen-permeable films
- Oxygen generators
- Validation of cold chain

Effects on Nutrients

Vitamin Content of Frozen Foods

- Vitamin C: Usually lost in a higher concentration than any other vitamin. A study was performed on peas to determine the cause of vitamin C loss. A vitamin loss of ten percent occurred during the blanching phase with the rest of the loss occurring during the cooling and washing stages. The vitamin loss was not actually accredited to the freezing process. Another experiment was performed involving peas and lima beans. Frozen and canned vegetables were both used in the experiment. The frozen vegetables were stored at −23 °C (−10 °F) and the canned vegetables were stored at room temperature 24 °C (75 °F). After 0, 3, 6, and 12 months of storage, the vegetables were analyzed with and without cooking. O'Hara, the scientist performing the experiment said, "From the view point of the vitamin content of the two vegetables when they were ready for the plate of the consumer, there did not appear to be any marked advantages attributable to method of preservation, frozen storage, processed in a tin, or processed in glass."
- Vitamin B_1 (Thiamin): A vitamin loss of 25 percent is normal. Thiamin is easily soluble in water and is destroyed by heat.
- Vitamin B_2 (Riboflavin): Not much research has been done to see how much freezing affects Riboflavin levels. Studies that have been performed are inconclusive; one study found an 18 percent vitamin loss in green vegetables, while another determined a 4 percent loss. It is commonly accepted that the loss of Riboflavin has to do with the preparation for freezing rather than the actual freezing process itself.
- Vitamin A (Carotene): There is little loss of carotene during preparation for freezing and freezing of most vegetables. Much of the vitamin loss is incurred during the extended storage period.

Effectiveness

Freezing is an effective form of food preservation because the pathogens that cause food spoilage are killed or do not grow very rapidly at reduced temperatures. The process is less effective in food preservation than are thermal techniques, such as boiling, because pathogens are more likely to be able to survive cold temperatures rather than hot temperatures. One of the problems surrounding the use of freezing as a method of food preservation is the danger that pathogens deactivated (but not killed) by the process will once again become active when the frozen food thaws.

A frozen food warehouse at McMurdo Station, Antarctica

Foods may be preserved for several months by freezing. Long-term frozen storage requires a constant temperature of –18 °C (0 °F) or less.

Quality

The speed of the freezing has a direct impact on the size and the number of ice crystals formed within a food product's cells and extracellular space. Slow freezing leads to fewer but larger ice crystals while fast freezing leads to smaller but more numerous ice crystals. Large ice crystals can puncture the walls of the cells of the food product which will cause a degradation of the texture of the product as well as the loss of its natural juices during thawing. That is why there will be a qualitative difference observed between food products frozen by ventilated mechanical freezing, non-ventilated mechanical freezing or cryogenic freezing with liquid nitrogen.

Reaction

According to a study, an American consumes on average 71 frozen foods a year, most of which are pre-cooked frozen meals. Many food critics host shows are dedicated to the tasting and reviewing of frozen foods, the web show *Freezerburns* being one of the more notable ones.

Curing (Food Preservation)

Curing is any of various food preservation and flavoring processes of foods such as meat, fish and vegetables, by the addition of combinations of salt, nitrates, nitrites, or sugar, with the aim of drawing moisture out of the food by the process of osmosis. Many curing processes also involve smoking, spicing, or cooking. Dehydration was the earliest form of food curing. Because curing increases the solute concentration in the food and hence decreases its water potential, the food becomes inhospitable for the microbe growth that causes food spoilage. Curing can be traced back to antiquity, and was the primary way of preserving meat and fish until the late 19th century.

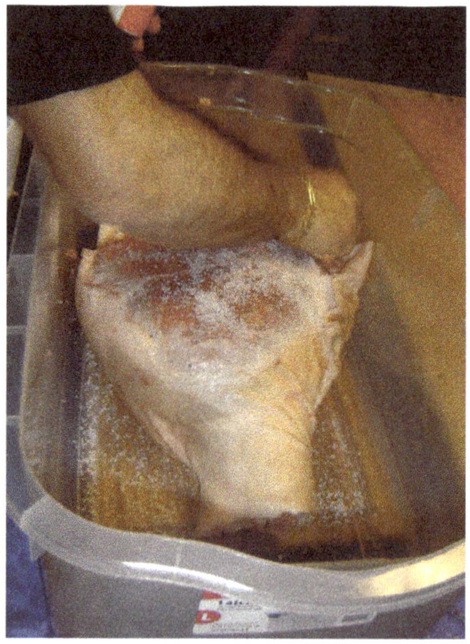

Sea salt being added to raw ham to make prosciutto

Nitrates and nitrites, in conjunction with salt, are one of the most common agents in curing meat because they further inhibit the growth of *Clostridium botulinum*. They also contribute to the characteristic pink color of cured meats like ham.

Meat preservation in general (of livestock, game, and poultry), is the set of all treatment processes for preserving the nutritious properties, taste, texture, and color of raw, partially cooked, or cooked meats while keeping them edible and safe to consume. Curing has been the dominant method of meat preservation for thousands of years, although modern developments like refrigeration and synthetic preservatives are now beginning to complement and supplant it.

Curing salt, also known as "Prague powder" or "pink salt", is typically a combination of salt and sodium nitrite that is dyed pink to distinguish it from ordinary salt.

Slices of beef in a box.

While meat preservation processes like curing were mainly developed in order to prevent disease and increase food security, the advent of modern preservation methods mean that in most developed countries today, curing is instead mainly practised for its cultural value and desirable impact on the texture and taste of food. For lesser-developed countries, curing remains a key process in ensuring the viability of meat production, transport and access.

Necessity of Curing

Untreated meat decomposes rapidly if it is not preserved, at a speed that depends on several factors, including ambient humidity, temperature, and the presence of pathogens. Most meats cannot be kept at room temperature in excess of a few days without spoiling, even in winter.

If kept in excess of this time, meat begins to change colour and exude a foul odour, indicating the decomposition of the food. Ingestion of such spoiled meat can cause serious food poisonings, like botulism.

While the short shelf life of fresh meat does not pose a significant problem when access to it is easy and supply is abundant, in times of scarcity and famine, or when the meat must be carried over long voyages, it spoils very quickly. In such circumstances the ability to preserve foods containing nutritional value for transport and storage is obvious.

Curing is able to significantly extend the life of meat before it spoils, by making it inhospitable to the growth of spoilage microbes.

History

A survival technique since prehistory, the conservation of meat has become, over the centuries, a topic of political, economic, and social importance worldwide.

Traditional Methods

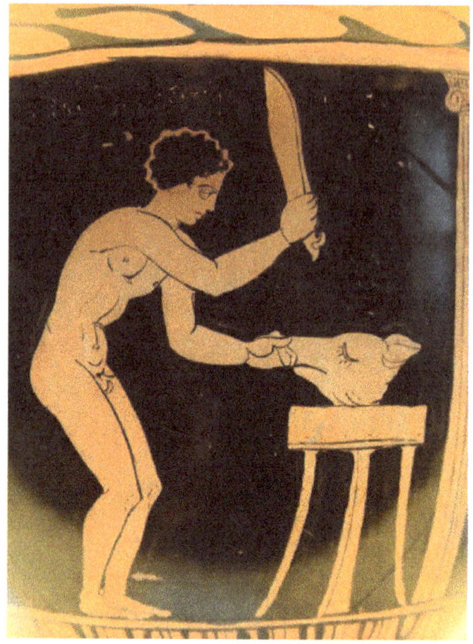

Young man preparing a pig's head after a sacrifice. Vase v. 360-340 BC, National Archaeological Museum of Spain

Food curing dates back to ancient times, both in the form of smoked meat and salt-cured meat.

Several sources describe the salting of meat in the ancient Mediterranean world. Diodore of Sicily in his *Bibliotheca historica* wrote that the Cosséens in the mountains of Persia salted the flesh of carnivorous animals. Strabo indicates that people at Borsippa were catching bats and salting them to eat. The ancient prepared *tarichos*, which was meat and fish conserved by salt or other means. The Romans called this dish *salsamentum* – which term later included salted fat, the sauces and spices used for its preparation. Also evidence of ancient sausage production exists. The Roman gourmet Apicius speaks of a sausage-making technique involving *œnogaros* (a mixture of the fermented fish sauce garum with oil or wine). Preserved meats were furthermore a part of religious traditions: resulting meat for offerings to the gods was salted before being given to priests, after which it could be picked up again by the offerer, or even sold in the butcher's.

A trade in salt meat occurred across ancient Europe. In Polybius's time, the Gauls exported salt pork each year to Rome in large quantities, where it was sold in different cuts: rear cuts, middle cuts, hams, and sausages. This meat, after having been salted with the greatest care, was sometime smoked. These goods had to have been considerably important, since they fed part of the Roman people and the armies. The Belgians were celebrated above all for the care which they gave to the fattening of their pigs. Their herds of sheep and pigs were so many, they could provide skins and salt meat not only for Rome, but also for most of Italy. The Ceretani of Spain drew a large export income from their hams, which were so succulent, they were in no way inferior to those of Cantabria. These *tarichos* of pig would become especially sought, to the point that the ancients considered this meat the most nourishing of all and the easiest to digest.

In Ethiopia, according to Pliny, and in Libya according to Saint Jerome, the Acridophages (literally, the locust-eaters) salted and smoked the crickets which arrived at their settlements in the spring in great swarms and which constituted, it was said, their sole food.

The smoking of meat was a traditional practice in North America, where Plains Indians hung their meat at the top of their tipis to increase the amount of smoke coming into contact with the food.

Middle Ages

In Europe, medieval cuisine made great use of meat and vegetables, and the guild of butchers was amongst the most powerful. During the 12th century, salt beef was consumed by all social classes. Smoked meat was called *carbouclée* in Romance tongues and *bacon* if it was pork.

The Middle Ages made pâté a masterpiece: that which is, in the 21st century, merely spiced minced meat (or fish), baked in a terrine and eaten cold, was at that time composed of a dough envelope stuffed with varied meats and superbly decorated for ceremonial feasts. The first French recipe, written in verse by Gace de La Bigne, mentions in the same pâté three great partridges, six fat quail, and a dozen larks. *Le Ménagier de Paris* mentions pâtés of fish, game, young rabbit, fresh venison, beef, pigeons, mutton, veal, and pork, and even pâtés of lark, turtledove, cow, baby bird, goose, and hen. Bartolomeo Sacchi, called Platine, prefect of the Vatican Library, gives the recipe for a pâté of wild beasts: the flesh, after being boiled with salt and vinegar, was larded and placed inside an envelope of spiced fat, with a *mélange* of pepper, cinnamon and pounded lard; one studded the fat with cloves until it was entirely covered, then placed it inside a pâte.

In the 16th century, the most fashionable pâtés were of woodcock, au bec doré, chapon, beef tongue, cow feet, sheep feet, chicken, veal, and venison. In the same era, Pierre Belon notes that the inhabitants of Crete and Chios lightly salted then oven-dried entire hares, sheep, and roe deer cut into pieces, and that in Turkey, cattle and sheep, cut and minced rouelles, salted then dried, were eaten on voyages with onions and no other preparation.

Early Modern Era

Barrels of salt beef in a reconstruction of an American Civil War stockpile, at Fort Macon State Park, North Carolina

During the Age of Discovery, salt meat was one of the main foods for sailors on long voyages, for instance in the merchant marine or the navy. In the 18th century, salted Irish beef, transported in barrels, was considered finest.

Scientific research on meat by chemists and pharmacists led to the creation of a new, extremely practical product: meat extract, which could appear in different forms. The need to properly feed soldiers during long campaigns outside the country, such as the Napoleonic Wars, and to nourish a constantly growing population often living in appalling conditions drove scientific research, but a confectioner, Nicolas Appert, in 1795 developed through experimentation a method which would become universal and in one language bears his name: airtight storage, called *appertisation* in French.

With the spread of *appertisation*, the 19th-century world entered the era of the "food industry", which developed new products such as canned salt meat (for example corned beef), but also led to lowered standards of food quality and hygiene – such as those Upton Sinclair described in *The Jungle*. These bad practices led to the creation of the Pure Food and Drug Act in 1906, followed by the national agencies for health security and the establishment of food traceability over the course of the 20th century. It also led to continuing technological innovation.

In France, the summer of 1857 was so hot that most butchers refused to slaughter animals and charcutiers lost considerable amounts of meat, due to inadequate conservation methods. A member of the Academy of Medicine and his son issued a 34-page summary of works printed between 1663 and 1857, which proposed some solutions: not less than 91 texts exist, of which 64 edited for only the years between 1851 and 1857.

Chemical Actions

Salt

Table salt (sodium chloride) is the primary ingredient used in meat curing. Removal of water and addition of salt to meat creates a solute-rich environment where osmotic pressure draws water out of microorganisms, slowing down their growth. Doing this requires a concentration of salt of nearly 20%. In addition, salt causes the soluble meat proteins to come to the surface of the meat particles within sausages. These proteins coagulate when the sausage is heated, helping to hold the sausage together. Finally, salt slows the oxidation process, effectively preventing the meat from going rancid.

Sugar

The sugar added to meat for the purpose of curing it comes in many forms, including honey, corn syrup solids, and maple syrup. However, with the exception of bacon, it does not contribute much to the flavor, but it does alleviate the harsh flavor of the salt. Sugar also contributes to the growth of beneficial bacteria such as *Lactobacillus* by feeding them.

Nitrates and Nitrites

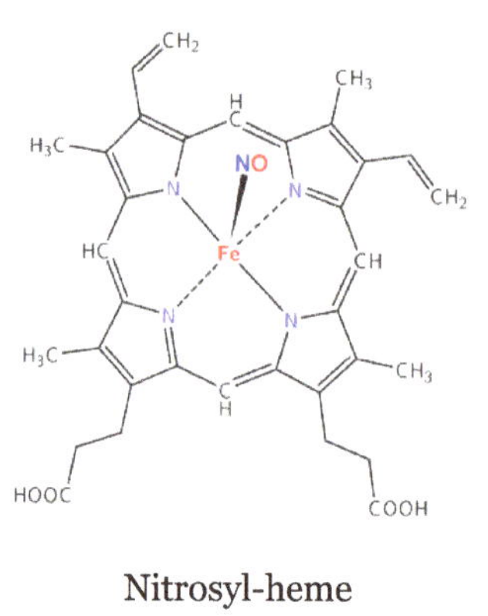

Nitrosyl-heme

Nitrosyl-heme

Nitrates and nitrites not only help kill bacteria, but also produce a characteristic flavor and give meat a pink or red color. Nitrite (NO−2) is generally supplied by sodium nitrite or (indirectly) by potassium nitrate. Nitrite salts are most often used in curing. Nitrate is specifically used only in a few curing conditions and products where nitrite (which may be generated from nitrate) must be generated in the product over long periods of time.

Nitrite further breaks down in the meat into nitric oxide (NO), which then binds to the iron atom in the center of myoglobin's heme group, reducing oxidation and causing a reddish-brown color (nitrosomyoglobin) when raw, and the characteristic cooked-ham pink color (nitrosohemochrome or nitrosyl-heme) when cooked. The addition of ascorbate to cured meat reduces formation of nitrosamines, but increases the nitrosylation of iron.

The use of nitrite and nitrate salts for meat curing goes back to the Middle Ages, and in the US has been formally used since 1925. Because of the relatively high toxicity of nitrite (the lethal dose in humans is about 22 mg/kg of body weight), the maximum allowed nitrite concentration in meat products is 200 ppm. Plasma nitrite is reduced in persons with endothelial dysfunction.

The use of nitrites in food preservation is controversial due to the potential for the formation of nitrosamines when nitrites are present in high concentrations and the product is cooked at high temperatures. The effect is seen for red or processed meat, but not for white meat or fish. The production of carcinogenic nitrosamines can be potently inhibited by the use of the antioxidants Vitamin C and the alpha-tocopherol form of Vitamin E during curing. Under simulated gastric conditions, nitrosothiols rather than nitrosamines are the main nitroso species being formed. The use of either compound is therefore regulated; for example, in the United States, the concentration of nitrates and nitrites is generally limited to 200 ppm or lower. They are considered irreplaceable in the prevention of botulism from consumption of cured dry sausages by preventing spore germination.

Smoke

Meat can also be preserved by "smoking", which means exposing it to smoke from burning or smoldering plant materials, usually wood. If the smoke is hot enough to slow-cook the meat, it will also keep it tender. One method of smoking calls for a smokehouse with damp wood chips or sawdust. In North America, hardwoods such as hickory, mesquite, and maple are commonly used for smoking, as are the wood from fruit trees such as apple, cherry, and plum, and even corncobs.

Smoking helps seal the outer layer of the food being cured, making it more difficult for bacteria to enter. It can be done in combination with other curing methods such as salting. Common smoking styles include hot smoking, smoke roasting (pit barbecuing) and cold smoking. Smoke roasting and hot smoking cook the meat while cold smoking does not. If the meat is cold smoked, it should be dried quickly to limit bacterial growth during the critical period where the meat is not yet dry. This can be achieved, as with jerky, by slicing the meat thinly.

Effect of Meat Preservation

On Health

The preservation of meat has allowed safe and plentiful access to this nutrient-rich food for thousands of years. It has therefore prevented countless cases of malnutrition and of food poisoning.

Since the 20th century, with respect to the relationship between diet and human disease (e.g. cardiovascular, etc.), scientists have conducted studies on the effects of lipolysis on vacuum-packed or frozen meat. In particular, by analyzing entrecôtes of frozen beef during 270 days at −20 °C (−4 °F), scientists found an important phospholipase that accompanies the loss of some unsaturated fat n-3 and n-6, which are already low in the flesh of ruminants.

In 2015, the International Agency for Research on Cancer of the World Health Organization classified processed meat, that is, meat that has undergone salting, curing, fermenting, and smoking, as "carcinogenic to humans".

On Trade

The improvement of methods of meat preservation, and of the means of transport of preserved products, has notably permitted the separation of areas of production and areas of consumption, which can now be distant without it posing a problem, permitting the exportation of meats.

For example, the appearance in the 1980s of preservation techniques under controlled atmosphere sparked a small revolution in the world's market for sheep meat: the lamb of New Zealand, one of the world's largest exporters of lamb, could henceforth be sold as fresh meat, since it could be preserved from 12 to 16 weeks, which would be a sufficient duration for it to reach Europe by boat. Before, meat from New Zealand was frozen, thus had a much lower value on European shelves. With the arrival of the new "chilled" meats, New Zealand could compete even more strongly with local producers of fresh meat. The use of controlled atmosphere to avoid the depreciation which affects frozen meat is equally useful in other meat markets, such as that for pork, which now also enjoys an international trade.

Smoking (Cooking)

Pork ribs being smoked

A Montreal Smoked Meat Sandwich, a well-known Canadian dish

Hot-smoked chum salmon

Smoking is the process of flavoring, cooking, or preserving food by exposing it to smoke from burning or smoldering material, most often wood. Meats and fish are the most common smoked foods, though cheeses, vegetables, and ingredients used to make beverages such as whisky, smoked beer, and *lapsang souchong* tea are also smoked.

In Europe, alder is the traditional smoking wood, but oak is more often used now, and beech to a lesser extent. In North America, hickory, mesquite, oak, pecan, alder, maple, and fruit-tree woods, such as apple, cherry, and plum, are commonly used for smoking. Other fuels besides wood can also be employed, sometimes with the addition of flavoring ingredients. Chinese tea-smoking uses a mixture of uncooked rice, sugar, and tea, heated at the base of a wok. Some North American ham and bacon makers smoke their products over burning corncobs. Peat is burned to dry and smoke the barley malt used to make whisky and some beers. In New Zealand, sawdust from the native manuka (tea tree) is commonly used for hot smoking fish. In Iceland, dried sheep dung is used to cold-smoke fish, lamb, mutton, and whale.

Historically, farms in the Western world included a small building termed the *smokehouse*, where meats could be smoked and stored. This was generally well-separated from other buildings both because of the fire danger and because of the smoke emanations.

Smoking of food is known to contaminate the food with carcinogenic polycyclic aromatic hydrocarbons.

History

The smoking of food dates back to the time of primitive cavemen. As caves or simple huts lacked chimneys, these dwellings could become very smoky. The early men would often hang meat up to dry, and they soon became aware that meat that was stored in smoky areas acquired a different flavor and was better preserved than meat that simply dried out. Over time this process was combined with pre-curing the food in salt or salty brines, resulting in a remarkably effective preservation process that was adapted and developed by numerous cultures around the world. Until the modern era, smoking was of a more "heavy duty" nature as the main goal was to preserve the food. Large quantities of salt were used in the curing process and smoking times were quite long, sometimes involving days of exposure.

The advent of modern transportation made it easier to transport food products over long distances and the need for the time and material intensive heavy salting and smoking declined. Smoking became more of a way to flavor than to preserve food. In 1939 a device called the *Torry Kiln* was invented at the Torry Research Station in Scotland. The kiln allowed for uniform mass-smoking and is considered the prototype for all modern large-scale commercial smokers. Although refinements in technique and advancements in technology have made smoking much easier, the basic steps involved remain essentially the same today as they were hundreds if not thousands of years ago.

Types

A "Little Chief" home smoker and racks with hot smoked Pacific halibut

- Cold smoking can be used as a flavor enhancer for items such as chicken breasts, beef, pork chops, salmon, scallops, and steak. The item is hung first to develop a pellicle, then can be cold smoked for just long enough to give some flavor. Some cold smoked foods are baked, grilled, steamed, roasted, or sautéed before eating. Smokehouse temperatures for cold smoking are typically done between 20 to 30 °C (68 to 86 °F). In this temperature range, foods take on a smoked flavor, but remain relatively moist. Cold smoking does not cook foods. Meats should be fully cured before cold smoking.
- Hot smoking exposes the foods to smoke and heat in a controlled environment. Like cold smoking, the item is hung first to develop a pellicle, then smoked. Although foods that have been hot smoked are often reheated or cooked, they are typically safe to eat without further cooking. Hams and ham hocks are fully cooked once they are properly smoked. Hot smoking occurs within the range of 52 to 80 °C (126 to 176 °F). Within this temperature range, foods are fully cooked, moist, and flavorful. If the smoker is allowed to get hotter than 185 °F (85 °C), the foods will shrink excessively, buckle, or even split. Smoking at high temperatures also reduces yield, as both moisture and fat are "cooked" away.
- Smoke roasting or smoke baking refers to any process that has the attributes of smoking combined with either roasting or baking. In North America, this smoking method is commonly referred to as "barbecuing", "pit baking", or "pit roasting". It may be done in a smoke roaster, closed wood-fired masonry oven or barbecue pit, any smoker that can reach above 250 °F (121 °C), or in a conventional oven by placing a pan filled with hardwood chips on the floor of the oven so the chips smolder and produce a smokebath. However, this should only be done in a well-ventilated area to prevent carbon monoxide poisoning.

Wood Smoke

Hickory-smoked country-style ribs

Hardwoods are made up mostly of three materials: cellulose, hemicellulose, and lignin. Cellulose and hemicellulose are the basic structural material of the wood cells; lignin acts as a kind of cell-bonding glue. Some softwoods, especially pines and firs, hold significant quantities of resin, which produces a harsh-tasting soot when burned; these woods are not often used for smoking.

Cellulose and hemicellulose are aggregate sugar molecules; when burnt, they effectively caramelize, producing carbonyls, which provide most of the color components and sweet, flowery, and fruity aromas. Lignin, a highly complex arrangement of interlocked phenolic molecules, also produces a number of distinctive aromatic elements when burnt, including smoky, spicy, and pungent compounds such as guaiacol, phenol, and syringol, and sweeter scents such as the vanilla-scented vanillin and clove-like isoeugenol. Guaiacol is the phenolic compound most responsible for the "smokey" taste, while syringol is the primary contributor to smokey aroma. Wood also contains small quantities of proteins, which contribute roasted flavors. Many of the odor compounds in wood smoke, especially the phenolic compounds, are unstable, dissipating after a few weeks or months.

A number of wood smoke compounds act as preservatives. Phenol and other phenolic compounds in wood smoke are both antioxidants, which slow rancidification of animal fats, and antimicrobials, which slow bacterial growth. Other antimicrobials in wood smoke include formaldehyde, acetic acid, and other organic acids, which give wood smoke a low pH—about 2.5. Some of these compounds are toxic to people as well, and may have health effects in the quantities found in cooking applications.

Since different species of trees have different ratios of components, various types of wood do impart a different flavor to food. Another important factor is the temperature at which the wood burns. High-temperature fires see the flavor molecules broken down further into unpleasant or flavorless compounds. The optimal conditions for smoke flavor are low, smoldering temperatures between 570 and 750 °F (299 and 399 °C). This is the temperature of the burning wood itself, not of the smoking environment, which uses much lower temperatures. Woods that are high in lignin content tend to burn hot; to keep them smoldering requires restricted oxygen supplies or a high moisture content. When smoking using wood chips or chunks, the combustion temperature is often raised by soaking the pieces in water before placing them on a fire.

Types of Smokers

There are a few basic types of smoker designs, each with their own advantages and disadvantages.

Charcoal Smokers

Offset Smokers

An example of a common offset smoker

The main characteristics of the offset smoker are that the cooking chamber is usually cylindrical in shape, with a shorter, smaller diameter cylinder attached to the bottom of one end for a firebox. To cook the meat, a small fire is lit in the firebox, where airflow is tightly controlled. The heat and smoke from the fire is drawn through a connecting pipe or opening into the cooking chamber. The heat and smoke cook and flavor the meat before escaping through an exhaust vent at the opposite end of the cooking chamber. Most manufacturers' models are based on this simple but effective design, and this is what most people picture when they think of a "BBQ smoker." Even large capacity commercial units use this same basic design of a separate, smaller fire box and a larger cooking chamber.

Upright Drum Smoker

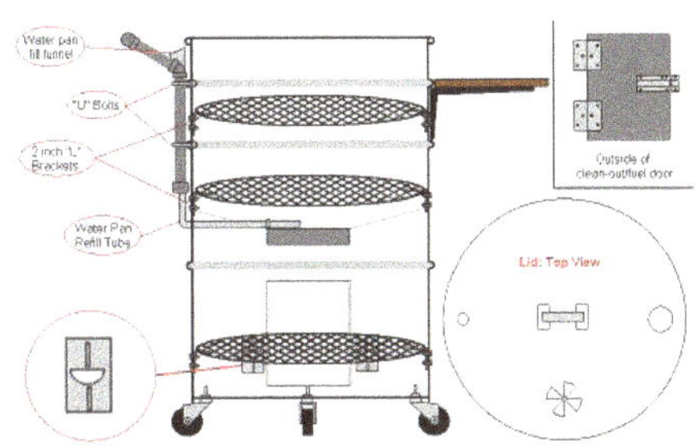

A diagram of a typical upright drum smoker

The upright drum smoker (also referred to as an ugly drum smoker or UDS) is exactly what its name suggests; an upright steel drum that has been modified for the purpose of pseudo-indirect hot smoking. There are many ways to accomplish this, but the basics include the use of a complete steel drum, a basket to hold charcoal near the bottom, and cooking rack (or racks) near the top; all covered by a vented lid of some sort. They have been built using many different sizes of steel drums, such as 30 US gallons (110 l; 25 imp gal), 55 US gallons (210 l; 46 imp gal), and 85 US gallons (320 l; 71 imp gal) for example, but the most popular size is the common 55 gallon drum.

This design is similar to smoking with indirect heat due to the distance from the coals and the racks (typically 24 inches (61 cm)). The temperatures used for smoking are controlled by limiting the amount of air intake at the bottom of the drum, and allowing a similar amount of exhaust out of vents in the lid. UDSs are very efficient with fuel consumption and flexible in their abilities to produce proper smoking conditions, with or without the use of a water pan or drip pan. Most UDS builders/users would say a water pan defeats the true pit BBQ nature of the UDS, as the drippings from the smoked meat should land on the coals, burning up, and imparting a unique flavor one cannot get with a water pan.

Vertical Water Smoker

A vertical water smoker (also referred to as a bullet smoker because of its shape) is a variation of the upright drum smoker. It uses charcoal or wood to generate smoke and heat, and contains a water bowl between the fire and the cooking grates. The water bowl serves to maintain optimal smoking temperatures and also adds humidity to the smoke chamber. It also creates an effect in which the water vapor and smoke condense together, which adds flavor to smoked foods. In addition, the bowl catches any drippings from the meat that may cause a flare-up. Vertical water smokers are extremely temperature stable and require very little adjustment once the desired temperature has been reached. Because of their relatively low cost and stable temperature, they are sometimes used in barbecue competitions where propane and electric smokers are not allowed.

A typical vertical water smoker

Propane Smoker

A diagram of a propane smoker, loaded with country style ribs and pork loin in foil.

A propane smoker is designed to allow the smoking of meat in a somewhat more controlled environment. The primary differences are the sources of heat and of the smoke. In a propane smoker, the heat is generated by a gas burner directly under a steel or iron box containing the wood or charcoal that provides the smoke. The steel box has few vent holes, on the top of the box only. By starving the heated wood of oxygen, it smokes instead of burning. Any combination of woods and charcoal may used. This method uses less wood.

Smoke Box Method

This more traditional method uses a two-box system: a fire box and a food box. The fire box is typically adjacent or under the cooking box, and can be controlled to a finer degree. The heat and smoke from the fire box exhausts into the food box, where it is used to cook and smoke the meat. These may be as simple as an electric heating element with a pan of wood chips placed on it, although more advanced models have finer temperature controls.

Commercial Smoke House

Commercial smokehouses, mostly made from stainless steel, have independent systems for smoke generation and cooking. Smoke generators use friction, an electric coil or a small flame to ignite sawdust on demand. Heat from steam coils or gas flames is balanced with live steam or water sprays to control the temperature and humidity. Elaborate air handling systems reduce hot or cold spots, to reduce variation in the finished product. Racks on wheels or rails are used to hold the product and facilitate movement.

Preservation

Smoked omul fish, endemic to Lake Baikal in Russia, on sale at Listyanka market

Smoke is an antimicrobial and antioxidant, but smoke alone is insufficient for preserving food in practice, unless combined with another preservation method. The main problem is the smoke compounds adhere only to the outer surfaces of the food; smoke does not actually penetrate far into meat or fish. In modern times, almost all smoking is carried out for its flavor. Artificial smoke flavoring can be purchased as a liquid to mimic the flavor of smoking, but not its preservative qualities.

In the past, smoking was a useful preservation tool, in combination with other techniques, most commonly salt-curing or drying. In some cases, particularly in climates without much hot sunshine, smoking was simply an unavoidable side effect of drying over a fire. For some long-smoked foods, the smoking time also served to dry the food. Drying, curing, or other techniques can render the interior of foods inhospitable to bacterial life, while the smoking gives the vulnerable exterior surfaces an extra layer of protection. For oily fish smoking is especially useful, as its antioxidant properties delay surface fat rancidification. (Interior fat is not as exposed to oxygen, which is what causes rancidity.) Some heavily-salted, long-smoked fish can keep without refrigeration for weeks or months. Such heavily-preserved foods usually require a treatment such as boiling in fresh water to make them palatable before eating.

Some Smoked Foods

Smoked Gruyère cheese

Fish being smoked in Tanji, The Gambia

Pastrami, a smoked and cured beef product

Many foods can be smoked. Some of the more common are listed here.

Beverages

- *Lapsang souchong* tea leaves are smoked and dried over pine or cedar fires

- Malt beverages
    - The malt used to make whisky
    - Rauchbier (smoked beer)

Fruit and Vegetables

- Capsicums: chipotles (smoked, ripe jalapeños), paprika
- Prunes (dried plums) can be smoked while drying
- *Wumei* are smoked plum fruits
- *Iburi-gakko* are a smoked daikon pickle from Akita Prefecture, Japan

Protein

- Cheeses
- Egg (eggs and fish eggs)
- Fish
    - Eel popular in eastern/northern Europe
    - Traditional Grimsby smoked fish (cod and haddock)
    - Haddock and Arbroath Smokies (haddock)
    - Kippers and bloater (herring)
    - Salmon
    - Mackerel
- Beef
    - Traditionally-prepared jerky
    - Pastrami (pickled, spiced and smoked beef brisket)
- Pork
    - Bacon
    - Ham
    - Bakkwa
    - Various sausages
- Tofu
- Turkey
- Various kinds of nuts

Spices

- Paprika
- Salt

Pickling

Pickling is the process of preserving or expanding the lifespan of food by either anaerobic fermentation in brine or immersion in vinegar. The resulting food is called a *pickle*, or, to prevent ambiguity, prefaced with the adjective *pickled*. The pickling procedure will typically affect the food's texture and flavor. In East Asia, vinaigrette (vegetable oil and vinegar) is also used as a pickling medium. Foods that are pickled include meats, fruits, eggs, and vegetables.

Another distinguishing characteristic is a pH 4.6 or lower, which is sufficient to kill most bacteria. Pickling can preserve perishable foods for months. Antimicrobial herbs and spices, such as mustard seed, garlic, cinnamon or cloves, are often added. If the food contains sufficient moisture, a pickling brine may be produced simply by adding dry salt. For example, German sauerkraut and Korean kimchi are produced by salting the vegetables to draw out excess water. Natural fermentation at room temperature, by lactic acid bacteria, produces the required acidity. Other pickles are made by placing vegetables in vinegar. Unlike the canning process, pickling (which includes fermentation) does not require that the food be completely sterile before it is sealed. The acidity or salinity of the solution, the temperature of fermentation, and the exclusion of oxygen determine which microorganisms dominate, and determine the flavor of the end product.

When both salt concentration and temperature are low, *Leuconostoc mesenteroides* dominates, producing a mix of acids, alcohol, and aroma compounds. At higher temperatures *Lactobacillus plantarum* dominates, which produces primarily lactic acid. Many pickles start with *Leuconostoc*, and change to *Lactobacillus* with higher acidity.

History

Pickling began 4000 years ago using cucumbers native to India. This was used as a way to preserve food for out-of-season use and for long journeys, especially by sea. Salt pork and salt beef were common staples for sailors before the days of steam engines. Although the process was invented to preserve foods, pickles are also made and eaten because people enjoy the resulting flavors. Pickling may also improve the nutritional value of food by introducing B vitamins produced by bacteria.

Etymology

The term *pickle* is derived from the Dutch word *pekel*, meaning *brine*. In the U.S. and Canada, and sometimes Australia and New Zealand, the word *pickle* alone almost always refers to a pickled cucumber, except when it is used figuratively. It may also refer to other types of pickles such as "pickled onion", "pickled cauliflower", etc. In the UK, pickle, as in a "cheese and pickle sandwich", may also refer to Ploughman's pickle, a kind of chutney.

Popularity of Pickles Around the World

Asia

South Asia

South Asia has a large variety of pickles (known as *achar* in Assamese, Punjabi, Hindi, Bengali, *uppinakaayi* in Kannada, *lonacha* in Marathi, *uppilittathu* or *achar* in Malayalam *oorukai* in Tamil, *ooragaya* in Telugu), which are mainly made from varieties of mango, lemon, lime, goongura(a sour leafy shrub), tamarind and Indian gooseberry (amla), chilli. Vegetables such as brinjal, carrots, cauliflower, tomato, bitter gourd, green tamarind, ginger, garlic, onion, and citron are also occasionally used. These fruits and vegetables are generally mixed with ingredients like salt, spices, and vegetable oils and are set to mature in a moistureless medium.

In Pakistan, pickles are known locally as *achaar* (in Urdu) and come in a variety of flavors. A popular item is the traditional mixed Hyderabadi pickle, a common delicacy prepared from an assortment of fruits (most notably mangoes) and vegetables blended with selected spices.

In Sri Lanka, *achcharu* is traditionally prepared from carrots, onions, and ground dates that are mixed with mustard powder, ground pepper, crushed ginger, garlic, and vinegar, and left to sit in a clay pot.

Southeast Asia

Singapore, Indonesian and Malaysian pickles, called *acar*, are typically made out of cucumber, carrot, bird's eye chilies, and shallots, these items being seasoned with vinegar, sugar and salt. Fruits, such as papaya and pineapple, are also sometimes pickled.

In the Philippines, *achara* is primarily made out of green papaya, carrots, and shallots, with cloves of garlic and vinegar. Other versions could include ginger, bell peppers, daikon, cucumbers or bamboo shoots. Separately, in some provinces, unripe mangoes or *burong mangga*, unripe tomatoes, green apple guavas, jicama turnips, bitter gourd and other fruit and vegetables are also pickled. Siling labuyo, sometimes with garlic and red onions, are also pickled in bottled vinegar. The spiced vinegar itself is a staple condiment in Filipino cuisine.

In Vietnam, vegetable pickles are called *dưa muối* ("salted vegetables") or *dưa chua* ("sour vegetables"). In Burma, tea leaves are pickled to produce lahpet, which has strong social and cultural importance.

Kimchi is a very common side dish in Korea.

East Asia

China is home to a huge variety of pickled vegetables, including radish, *baicai* (Chinese cabbage, notably *suan cai, la bai cai, pao cai,* and Tianjin preserved vegetable), *zha cai,* chili pepper, and cucumbers, among many others.

Japanese *tsukemono* (pickled foods) include *takuan* (daikon), *umeboshi* (ume plum), *gari* & *beni shoga* (ginger), turnip, cucumber, and Chinese cabbage.

The Korean staple kimchi is usually made from pickled napa cabbage and radish, but is also made from green onions, garlic stems, chives and a host of other vegetables. Kimchi is popular throughout East Asia. Jangajji is another example of pickled vegetables.

Western Asia

Torshi, traditional pickles in Southeast Europe, Western Asia and the Caucasus

In Iran, Turkey, Arab countries, the Balkans, and the Caucasus, pickles (called torshi in Persian, *turşu* in Turkish language and *mekhallel* in Arabic) are commonly made from turnips, peppers, carrots, green olives, cucumbers, cabbage, green tomatoes, lemons, and cauliflower.

Europe

Central and Eastern Europe

Coriander seeds are one of the spices popularly added to pickled vegetables in Europe.

In Hungary the main meal *(lunch)* usually goes with some kind of pickles *(savanyúság)* but they are commonly consumed at other times of the day too. The most commonly consumed pickles are sauerkraut *(savanyú káposzta)*, the different kinds of pickled cucumbers and peppers and *csalamádé* but tomatoes, carrots, beetroot, baby corn, onions, garlic, certain squashes and melons and a few fruits like plums and apples are used to make pickles too. Stuffed pickles are specialties usually made of peppers or melons pickled after being stuffed with a cabbage filling. Pickled plum stuffed with garlic is a unique Hungarian type of pickle just like *csalamádé* and leavened cucumber *(kovászos uborka)*. *Csalamádé* a type of mixed pickle made of cabbage, cucumber, paprika, onion, carrot, tomatoes and bay leaf mixed up with vinegar as the fermenting agent. Leavened cucumber, unlike other types of pickled cucumbers that are around all year long, is rather a seasonal pickle produced in the summer. Cucumbers, spices, herbs and slices of bread are put in a glass jar with salt water and kept in direct sunlight for a few days. The yeast from the bread, along with other pickling agents and spices fermented under the hot sun, give the cucumbers a unique flavor, texture and slight carbonation. Its juice can be used to make a special type of spritzer *('Újházy fröccs')* instead of carbonated water. It is common for Hungarian households to produce their own pickles. Different regions or towns have their special recipes unique to them. Among them all the Vecsési Sauerkraut *(Vecsési savanyú káposzta)* is the most famous.

Pickled tomatoes are common in CIS

Romanian pickles (murături) are made out of beetroot, cucumbers, green tomatoes (*gogonele*), carrots, cabbage, bell peppers, melons, mushrooms, turnips, celery and cauliflower. Meat, like pork, can also be preserved in salt and lard.

Polish and Czech traditional pickles are cucumbers and sauerkraut, but other pickled fruits and vegetables, including plums, pumpkins and mushrooms are also common.

Russian, Ukrainian and Belarusian pickled items include beets, mushrooms, tomatoes, sauerkraut, cucumbers, ramsons, garlic, eggplant (which is typically stuffed with julienned carrots), custard squash, and watermelon. In these countries countries garden produce is commonly pickled using salt, dill, blackcurrant leaves, bay leaves and garlic and is stored in a cool, dark place.

Southern Europe

An Italian pickled vegetable dish is giardiniera, which includes onions, carrots, celery and cauliflower. Many places in southern Italy, particularly in Sicily, pickle eggplants and hot peppers.

In Albania, Bulgaria, Serbia, Macedonia and Turkey, mixed pickles, known as *turshi* or *turshu* form popular appetizers, which are typically eaten with *rakia*. Pickled green tomatoes, cucumbers, carrots, bell peppers, peppers, eggplants, and sauerkraut are also popular.

Turkish pickles, called *turşu*, are made out of vegetables, roots, and fruits such as peppers, cucumber, Armenian cucumber, cabbage, tomato, eggplant (aubergine), carrot, turnip, beetroot, green almond, baby watermelon, baby cantaloupe, garlic, cauliflower, bean and green plum. A mixture of spices flavor the pickles.

In Greece, pickles are made out of carrots, celery, eggplants stuffed with diced carrots, cauliflower, tomatoes, and peppers.

Northern Europe

In Britain, pickled onions and pickled eggs are often sold in pubs and fish and chip shops. Pickled beetroot, walnuts, and gherkins, and condiments such as Branston Pickle and piccalilli are typically eaten as an accompaniment to pork pies and cold meats, sandwiches or a ploughman's lunch. Other popular pickles in the UK are pickled mussels, cockles, red cabbage, mango chutney, sauerkraut, and olives. Rollmops are also quite widely available under a range of names from various producers both within and without the UK.

Pickled herring, rollmops, and salmon are popular in Scandinavia. Pickled cucumbers and red garden beets are important as condiments for several traditional dishes. Pickled capers are also common in Scandinavian cuisine.

United States and Canada

A dish of giardiniera

In the United States and Canada, pickled cucumbers (most often referred to simply as "pickles" in

Canada and the United States), olives, and sauerkraut are most popular, although pickles popular in other nations are also available.

Canadian pickling is similar to that of Britain. Through the winter, pickling is an important method of food preservation. Pickled cucumbers, onions, and eggs are very popular. Chow Chow is a tart vegetable mix popular in the Maritime Provinces and the Southern United States, similar to picalily. Pickled fish is very popular, as in Scotland. Meat is often also pickled or preserved in different brines throughout the winter, most prominently in the harsh climate of Newfoundland.

Giardiniera, a mixture of pickled peppers, celery and olives, is a popular condiment in Chicago and other cities with large Italian-American populations, and is often consumed with Italian beef sandwiches. Pickled eggs are common in the Upper Peninsula of Michigan. Pickled herring is available in the Upper Midwest. Pennsylvania Dutch Country has a strong tradition of pickled foods, including chow-chow and red beet eggs. In the Southern United States, pickled okra and watermelon rind are popular, as are deep-fried pickles and pickled pig's feet, pickled chicken eggs, pickled quail eggs, pickled Garden vegetables and pickled sausage. In Mexico, chili peppers, particularly of the Jalapeño and serrano varieties, pickled with onions, carrots and herbs form common condiments. Various pickled vegetables, fish, or eggs may make a side dish to a Canadian lunch or dinner.

In the United States, National Pickle Day is a popular food holiday every year on November 14.

Mexico, Central America, and South America

In the Mesoamerican region pickling is known as "encurtido" or "curtido" for short. The pickles or "curtidos" as known in Latin America are served cold, as an appetizer, as a side dish or as a tapas dish in Spain. In several Central American countries it is prepared with cabbage, onions, carrots, lemon, vinegar, oregano, and salt. In Mexico, "curtido" consists of carrots, onions, and jalapeño peppers and used to accompany meals still common in taquerias and restaurants. In order to prepare a carrot "curtido" simply add carrots to vinegar and other ingredients that are common to the region such as chilli, tomato, and onions. Varies depending on the food, in the case of sour. Another example of a type of pickling which involves the pickling of meats or seafood is the "escabeche" or "ceviches" popular in Peru, Ecuador, and throughout Latin America and the Caribbean. These dishes include the pickling of pig's feet, pig's ears, and gizzards prepared as an "escabeche" with spices and seasonings to flavor it. The ceviches consists of shrimp, octopus, and various fishes seasoned and served cold.

Process

In chemical pickling, the jar and lid are first boiled in order to sterilize them. The fruits or vegetables to be pickled are then added to the jar along with brine, vinegar, or both, as well as spices, and are then allowed to ferment until the desired taste is obtained.

The food can be pre-soaked in brine before transferring to vinegar. This reduces the water content of the food which would otherwise dilute the vinegar. This method is particularly useful for fruit and vegetables with a high natural water content.

In commercial pickling, a preservative like sodium benzoate or EDTA may also be added to enhance shelf life. In fermentation pickling, the food itself produces the preservation agent, typically by a process involving Lactobacillus bacteria that produce lactic acid as the preservative agent.

Vase by Bát Tràng porcelain for pickling

Alum is used in pickling to promote crisp texture and approved as a food additive by the United States Food and Drug Administration.

"Refrigerator pickles" are unfermented pickles made by marinating fruit or vegetables in a seasoned vinegar solution. They must be stored under refrigeration or undergo canning to achieve long-term storage.

Health Benefits

Traditionally manufactured pickles are source of healthy probiotic microbes, which occur by natural fermentation in brine, but pickles produced using vinegar are not probiotic.

Possible Health Hazards of Pickled Vegetables

The *British Journal of Cancer* released an online 2009 meta-analysis of research on pickles as increasing the risks of esophageal cancer. The report, citing limited data in a statistical meta analysis, indicates a potential two-fold increased risk of oesophageal cancer associated with Asian pickled vegetable consumption. Results from the research are described as having "high heterogeneity" and the study said that further well-designed prospective studies were warranted.

Some common fungi can facilitate the formation of N-nitroso compounds, which are strong oesophageal carcinogens in several animal models. Roussin red methyl ester, a non-alkylating nitroso compound with tumour-promoting effect in vitro, was identified in pickles from Linxian in much higher concentrations than in samples from low-incidence areas. Fumonisin mycotoxins have been shown to cause liver and kidney tumours in rodents.

Canning

Canning is a method of preserving food in which the food contents are processed and sealed in an airtight container. Canning provides a shelf life typically ranging from one to five years, although under specific circumstances it can be much longer. A freeze-dried canned product, such as canned dried lentils, could last as long as 30 years in an edible state. In 1974, samples of canned food from the wreck of the *Bertrand*, a steamboat that sank in the Missouri River in 1865, were tested by the National Food Processors Association. Although appearance, smell and vitamin content had deteriorated, there was no trace of microbial growth and the 109-year-old food was determined to be still safe to eat.

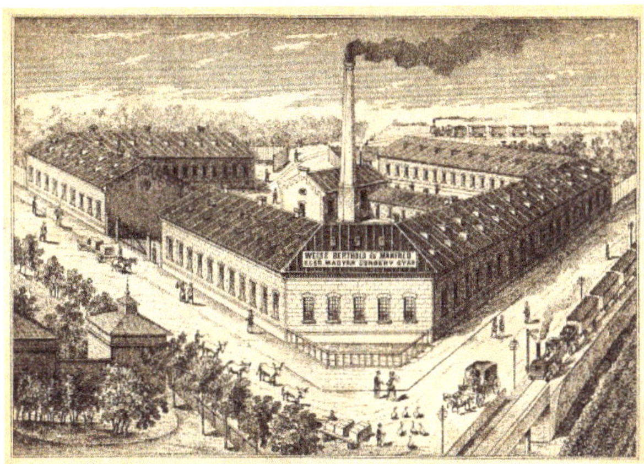

The Berthold-Weiss Factory, one of the first large canned food factories in Csepel-Budapest (1885)

Exemple d'installation d'une fabrique de conserves.

How canned food was made, picture from Albert Seigneurie's Grocery Encyclopedia (1898)

History and Development of Canning

French Origins

During the first years of the Napoleonic Wars, the French government offered a hefty cash award of 12,000 francs to any inventor who could devise a cheap and effective method of preserving large amounts of food. The larger armies of the period required increased and regular supplies of quality food. Limited food availability was among the factors limiting military campaigns to the summer and autumn months. In 1809, Nicolas Appert, a French confectioner and brewer, observed that food cooked inside a jar did not spoil unless the seals leaked, and developed a method of sealing food in glass jars. Appert was awarded the prize in 1810 by Count Montelivert, a French minister of the interior. The reason for lack of spoilage was unknown at the time, since it would be another 50 years before Louis Pasteur demonstrated the role of microbes in food spoilage.

Appert canning jar

Nicolas Appert, developer of the canning process

The French Army began experimenting with issuing canned foods to its soldiers, but the slow process of canning foods and the even slower development and transport stages prevented the army from shipping large amounts across the French Empire, and the war ended before the process was perfected. Unfortunately for Appert, the factory which he had built with his prize money was razed in 1814 by Allied soldiers invading France.

Following the end of the Napoleonic Wars, the canning process was gradually employed in other European countries and in the US.

In the United Kingdom

A Dixie Can Sealer for home use. Now in Thinktank, Birmingham Science Museum.

Based on Appert's methods of food preservation, the tin can process was allegedly developed by Frenchman Philippe de Girard, who came to London and used British merchant Peter Durand as an agent to patent his own idea in 1810. Durand did not pursue food canning himself, selling his patent in 1811 to Bryan Donkin and John Hall, who were in business as Donkin Hall and Gamble, of Bermondsey. Bryan Donkin developed the process of packaging food in sealed airtight cans, made of tinned wrought iron. Initially, the canning process was slow and labour-intensive, as each large can had to be hand-made, and took up to six hours to cook, making canned food too expensive for ordinary people.

The main market for the food at this stage was the British Army and Royal Navy. By 1817 Donkin recorded that he had sold £3000 worth of canned meat in six months. In 1824 Sir William Edward Parry took canned beef and pea soup with him on his voyage to the Arctic in HMS Fury, during his search for a northwestern passage to India. In 1829, Admiral Sir James Ross also took canned food to the Arctic, as did Sir John Franklin in 1845. Some of his stores were found by the search expedition led by Captain (later Admiral Sir) Leopold McLintock in 1857. One of these cans was opened in 1939, and was edible and nutritious, though it was not analysed for contamination by the lead solder used in its manufacture.

In Europe

Throughout the mid-19th century, canned food became a status symbol amongst middle-class households in Europe, being something of a frivolous novelty. Early methods of manufacture employed poisonous lead solder for sealing the cans, which may have worsened the disastrous outcome of the 1845 Franklin expedition to chart and navigate the Northwest Passage.

Increasing mechanisation of the canning process, coupled with a huge increase in urban populations across Europe, resulted in a rising demand for canned food. A number of inventions and improvements followed, and by the 1860s smaller machine-made steel cans were possible, and the time to cook food in sealed cans had been reduced from around six hours to thirty minutes.

In the United States

1914 magazine advertisement for cookware with instructions for home canning

Canned food also began to spread beyond Europe — Robert Ayars established the first American canning factory in New York City in 1812, using improved tin-plated wrought-iron cans for preserving oysters, meats, fruits and vegetables. Demand for canned food greatly increased during wars. Large-scale wars in the nineteenth century, such as the Crimean War, American Civil War, and Franco-Prussian War introduced increasing numbers of working-class men to canned food, and allowed canning companies to expand their businesses to meet military demands for non-perishable food, allowing companies to manufacture in bulk and sell to wider civilian markets after wars ended. Urban populations in Victorian Britain demanded ever-increasing quantities of cheap, varied, quality food that they could keep at home without having to go shopping daily. In response, companies such as Underwood, Nestlé, Heinz, and others provided quality canned food for sale to working class city-dwellers. In particular, Crosse and Blackwell took over the concern of Donkin Hall and Gamble. The late 19th century saw the range of canned food available to urban populations greatly increase, as canners competed with each other using novel foodstuffs, highly decorated printed labels, and lower prices.

World War I

Demand for canned food skyrocketed during World War I, as military commanders sought vast quantities of cheap, high-calorie food to feed their millions of soldiers, which could be transported safely, survive trench conditions, and not spoil in transport. Throughout the war, soldiers generally subsisted on low-quality canned foodstuffs, such as the British "Bully Beef" (cheap corned beef), pork and beans, canned sausages and Maconochies Irish Stew, but by 1916 widespread boredom with cheap canned food amongst soldiers resulted in militaries purchasing better-quality food to improve morale, and the complete meals in a can began to appear. In 1917 the French Army began issuing canned French cuisine, such as coq au vin, Beef Bourguignon and Vichyssoise while the Italian Army experimented with canned ravioli, spaghetti bolognese, Minestrone and Pasta e fagioli. Shortages of canned food in the British Army in 1917 led to the government issuing cigarettes and amphetamines to soldiers to suppress their appetites. After the war, companies that had supplied military canned food improved the quality of their goods for civilian sale.

Methods

The original fragile and heavy glass containers presented challenges for transportation, and glass jars were largely replaced in commercial canneries with cylindrical tin or wrought-iron canisters (later shortened to "cans") following the work of Peter Durand (1810). Cans are cheaper and quicker to make, and much less fragile than glass jars. Glass jars have remained popular for some high-value products and in home canning. Can openers were not invented for another thirty years — at first, soldiers had to cut the cans open with bayonets or smash them open with rocks. Today, tin-coated steel is the material most commonly used. Laminate vacuum pouches are also used for canning, such as used in MREs and Capri Sun drinks.

To prevent the food from being spoiled before and during containment, a number of methods are used: pasteurisation, boiling (and other applications of high temperature over a period of time), refrigeration, freezing, drying, vacuum treatment, antimicrobial agents that are natural to the recipe of the foods being preserved, a sufficient dose of ionizing radiation, submersion in a strong saline solution, acid, base, osmotically extreme (for example very sugary) or other microbially-challenging environments.

Other than sterilization, no method is perfectly dependable as a preservative. For example, the microorganism *Clostridium botulinum* (which causes botulism) can only be eliminated at temperatures above the boiling point of water.

From a public safety point of view, foods with low acidity (a pH more than 4.6) need sterilization under high temperature (116-130 °C). To achieve temperatures above the boiling point requires the use of a pressure canner. Foods that must be pressure canned include most vegetables, meat, seafood, poultry, and dairy products. The only foods that may be safely canned in an ordinary boiling water bath are highly acidic ones with a pH below 4.6, such as fruits, pickled vegetables, or other foods to which acidic additives have been added.

Double Seams

Invented in 1888 by Max Ams, modern double seams provide an airtight seal to the tin can. This airtight nature is crucial to keeping micro-organisms out of the can and keeping its contents sealed inside. Thus, double seamed cans are also known as Sanitary Cans. Developed in 1900 in Europe, this sort of can was made of the traditional cylindrical body made with tin plate. The two ends (lids) were attached using what is now called a double seam. A can thus sealed is impervious to contamination by creating two tight continuous folds between the can's cylindrical body and the lids. This eliminated the need for solder and allowed improvements in manufacturing speed, reducing cost.

Double seaming uses rollers to shape the can, lid and the final double seam. To make a sanitary can and lid suitable for double seaming, manufacture begins with a sheet of coated tin plate. To create the can body, rectangles are cut and curled around a die, and welded together creating a cylinder with a side seam.

Rollers are then used to flare out one or both ends of the cylinder to create a quarter circle flange around the circumference. Precision is required to ensure that the welded sides are perfectly aligned, as any misalignment will cause inconsistent flange shape, compromising its integrity.

A circle is then cut from the sheet using a die cutter. The circle is shaped in a stamping press to create a downward countersink to fit snugly into the can body. The result can be compared to an upside down and very flat top hat. The outer edge is then curled down and around about 140 degrees using rollers to create the end curl.

The result is a steel tube with a flanged edge, and a countersunk steel disc with a curled edge. A rubber compound is put inside the curl.

Seaming

Opened can

The body and end are brought together in a seamer and held in place by the base plate and chuck, respectively. The base plate provides a sure footing for the can body during the seaming operation and the chuck fits snugly into the end (lid). The result is the countersink of the end sits inside the top of the can body just below the flange. The end curl protrudes slightly beyond the flange.

First Operation

Once brought together in the seamer, the seaming head presses a first operation roller against the end curl. The end curl is pressed against the flange curling it in toward the body and under the flange. The flange is also bent downward, and the end and body are now loosely joined together. The first operation roller is then retracted. At this point five thicknesses of steel exist in the seam. From the outside in they are:

- End
- Flange
- End Curl
- Body
- Countersink

Second Operation

The seaming head then engages the second operation roller against the partly formed seam. The second operation presses all five steel components together tightly to form the final seal. The five layers in the final seam are then called; a) End, b) Body Hook, c) Cover Hook, d) Body, e) Countersink. All sanitary cans require a filling medium within the seam because otherwise the metal-to-metal contact will not maintain a hermetic seal. In most cases, a rubberized compound is placed inside the end curl radius, forming the critical seal between the end and the body.

Probably the most important innovation since the introduction of double seams is the welded side seam. Prior to the welded side seam, the can body was folded and/or soldered together, leaving a relatively thick side seam. The thick side seam required that the side seam end juncture at the end curl to have more metal to curl around before closing in behind the Body Hook or flange, with a greater opportunity for error.

Seamer Setup and Quality Assurance

Many different parts during the seaming process are critical in ensuring that a can is airtight and vacuum sealed. The dangers of a can that is not hermetically sealed are contamination by foreign objects (bacteria or fungicide sprays), or that the can could leak or spoil.

One important part is the seamer setup. This process is usually performed by an experienced technician. Amongst the parts that need setup are seamer rolls and chucks which have to be set in their exact position (using a feeler gauge or a clearance gauge). The lifter pressure and position, roll and chuck designs, tooling wear, and bearing wear all contribute to a good double seam.

Incorrect setups can be non-intuitive. For example, due to the springback effect, a seam can appear loose, when in reality it was closed too tight and has opened up like a spring. For this reason, experienced operators and good seamer setup are critical to ensure that double seams are properly closed.

Quality control usually involves taking full cans from the line - one per seamer head, at least once or twice per shift, and performing a teardown operation (wrinkle/tightness), mechanical tests (external thickness, seamer length/height and countersink) as well as cutting the seam open with a twin blade saw and measuring with a double seam inspection system. The combination of these measurements will determine the seam's quality.

Use of a Statistical Process Control (SPC) software in conjunction with a manual double-seam monitor, computerized double seam scanner, or even a fully automatic double seam inspection system makes the laborious process of double seam inspection faster and much more accurate. Statistically tracking the performance of each head or seaming station of the can seamer allows for better prediction of can seamer issues, and may be used to plan maintenance when convenient, rather than to simply react after bad or unsafe cans have been produced.

Nutritional Value

Canning is a way of processing food to extend its shelf life. The idea is to make food available and edible long after the processing time. A 1997 study found that canned fruits and vegetables are as rich with dietary fiber and vitamins as the same corresponding fresh or frozen foods, and in some cases the canned products are richer than their fresh or frozen counterparts. The heating process during canning appears to make dietary fiber more soluble, and therefore more readily fermented in the colon into gases and physiologically active byproducts. Canned tomatoes have a higher available lycopene content. Consequently, canned meat and vegetables are often among the list of food items that are stocked during emergencies. In 2013, the Can Manufacturers Institute launched the Cans Get You Cooking Campaign with the support of Crown Holdings, Inc., Ball Corporation, and Silgan Containers. The goal of the campaign is to get consumers to use more canned goods in their daily meals.

Potential Hazards

Women working in a cannery

In the beginning of the 19th century the process of canning foods was mainly done by small canneries. These canneries were full of overlooked sanitation problems, such as poor hygiene and unsanitary work environments. Since the refrigerator did not exist and industrial canning standards were not set in place it was very common for contaminated cans to slip onto the grocery store shelves.

Migration of Can Components

In canning toxicology, *migration* is the movement of substances from the can itself into the contents. Potential toxic substances that can migrate are lead, causing lead poisoning, or bisphenol A, a potential endocrine disruptor that is an ingredient in the epoxy commonly used to coat the inner surface of cans.

Salt Content

Salt (sodium chloride), dissolved in water, is used in the canning process. As a result, canned food can be a major source of dietary salt. Too much salt increases the risk of health problems, including high blood pressure. Therefore, health authorities have recommended limitations of dietary sodium. Many canned products are available in low-salt and no-salt alternatives.

Rinsing thoroughly after opening may reduce the amount of salt in canned foods, since much of the salt content is thought to be in the liquid, rather than the food itself.

Botulism

Foodborne botulism results from contaminated foodstuffs in which *C. botulinum* spores have been allowed to germinate and produce botulism toxin, and this typically occurs in canned non-acidic food substances. *C. botulinum* prefers low oxygen environments, and can therefore grow in canned foods. Botulism is a rare but serious paralytic illness, leading to paralysis that typically starts with the muscles of the face and then spreads towards the limbs. In severe forms, it leads to paralysis of the breathing muscles and causes respiratory failure. In view of this life-threatening complication, all suspected cases of botulism are treated as medical emergencies, and public health officials are usually involved to prevent further cases from the same source.

Canning and Economic Recession

Canned goods and canning supplies sell particularly well in times of recession due to the tendency of financially stressed individuals to engage in cocooning, a term used by retail analysts to describe the phenomenon in which people choose to stay at home instead of adding expenditures to their budget by dining out and socializing outside the home.

In February 2009 during a recession, the United States saw an 11.5% rise in sales of canning-related items.

Some communities in the US have county canning centers which are available for teaching canning, or shared community kitchens which can be rented for canning one's own foods.

In Popular Culture

There was a 1956 US documentary *The Miracle of the Can* that mentions the Pea Tenderomiter for the canning trade.

Notable Canned Foods

Harker & Badia canned tomatoes. Valencia, 1920s.

- Baked beans
- Canned Asparagus
- Canned fish
- Canned Ham
- Canned sardines
- Canned tuna
- Cream of mushroom soup
- Mushrooms
- Peaches
- Pears
- Pineapple
- Pink salmon
- Spam
- Tomato juice
- Tomato soup
- Vegetables - usually carrots, peas, maize and spinach
- Vienna sausages

Antioxidant

An antioxidant is a molecule that inhibits the oxidation of other molecules. Oxidation is a chemical reaction that can produce free radicals, leading to chain reactions that may damage cells. Antioxidants such as thiols or ascorbic acid (vitamin C) terminate these chain reactions. The term "antioxidant" is mainly used for two different groups of substances: industrial chemicals which are added to products to prevent oxidation, and natural chemicals found in foods and body tissue which are said to have beneficial health effects.

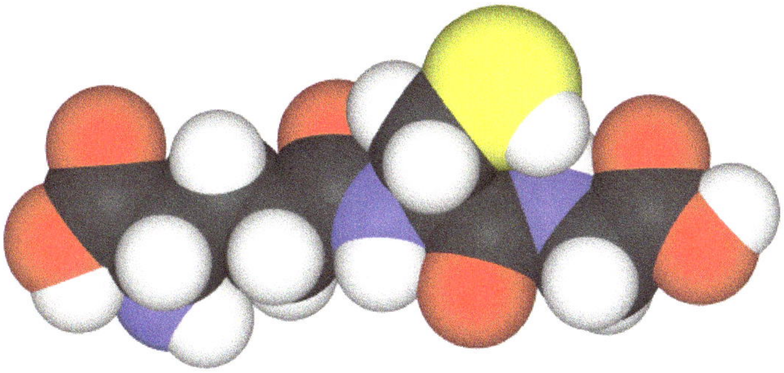

Model of the antioxidant metabolite glutathione. The yellow sphere is the redox-active sulfur atom that provides antioxidant activity, while the red, blue, white, and dark grey spheres represent oxygen, nitrogen, hydrogen, and carbon atoms, respectively.

To balance the oxidative state, plants and animals maintain complex systems of overlapping antioxidants, such as glutathione and enzymes (e.g., catalase and superoxide dismutase) produced internally or the dietary antioxidants, vitamin A, vitamin C and vitamin E.

Diets containing antioxidant dietary supplements do not improve health nor are they effective in preventing diseases. Randomized clinical trials including supplements of beta-carotene, vitamin A and vitamin E singly or in different combinations found no effect on mortality rate and cancer risk, or may even increase cancer risk. Supplementation with selenium or vitamin E does not reduce the risk of cardiovascular disease. Oxidative stress can be considered as either a cause or consequence of some diseases, an area of research stimulating drug development for antioxidant compounds for use as potential therapies.

Industrial antioxidants have diverse uses, such as food and cosmetics preservatives and inhibitors of rubber or gasoline deterioration.

Health Effects

Relation to Diet

Although certain levels of antioxidant vitamins in the diet are required for good health, there is considerable doubt as to whether antioxidant-rich foods or supplements have anti-disease activity; and if they are actually beneficial, it is unknown which antioxidant(s) are needed from the diet and in what amounts beyond typical dietary intake. Some authors dispute the hypothesis that antioxidant vitamins could prevent chronic diseases, while others maintain such a possibility is unproved and misguided from the beginning.

Polyphenols, which often have antioxidant properties in vitro, are not necessarily antioxidants in vivo due to extensive metabolism. In many polyphenols, the catechol group acts as electron acceptor and is therefore responsible for the antioxidant activity. However, this catechol group undergoes extensive metabolism upon uptake in the human body, for example by Catechol-O-methyl transferase, and is therefore no longer able to act as electron acceptor. Many polyphenols may have non-antioxidant roles in minute concentrations that affect cell-to-cell signaling, receptor sensitivity, inflammatory enzyme activity or gene regulation.

Although dietary antioxidants have been investigated for potential effects on neurodegenerative diseases such as Alzheimer's disease, Parkinson's disease, and amyotrophic lateral sclerosis, these studies have been inconclusive.

Drug Candidates

Tirilazad is an antioxidant steroid derivative that inhibits the lipid peroxidation that is believed to play a key role in neuronal death in stroke and head injury. It demonstrated activity in animal models of stroke, but human trials demonstrated no effect on mortality or other outcomes in subarachnoid haemorrhage and worsened results in ischemic stroke.

Similarly, the designed antioxidant NXY-059 exhibited efficacy in animal models, but failed to improve stroke outcomes in a clinical trial. As of November 2014, other antioxidants are being studied as potential neuroprotectants.

Common pharmaceuticals (and supplements) with antioxidant properties may interfere with the efficacy of certain anticancer medication and radiation.

Physical Exercise

During exercise, oxygen consumption can increase by a factor of more than 10. However, no benefits for physical performance to athletes are seen with vitamin E supplementation and 6 weeks of vitamin E supplementation had no effect on muscle damage in ultramarathon runners. Some research suggests that supplementation with amounts as high as 1000 mg of vitamin C inhibits recovery. Other studies indicated that antioxidant supplementation may attenuate the cardiovascular benefits of exercise.

Adverse Effects

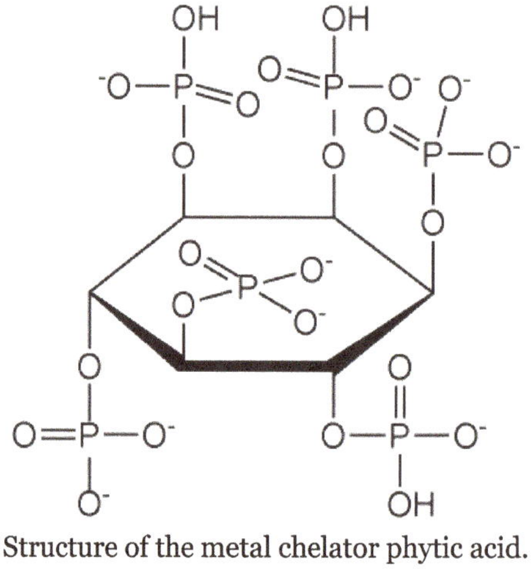

Structure of the metal chelator phytic acid.

Relatively strong reducing acids can have antinutrient effects by binding to dietary minerals such as iron and zinc in the gastrointestinal tract and preventing them from being absorbed. Notable examples are oxalic acid, tannins and phytic acid, which are high in plant-based diets. Calcium and iron deficiencies are not uncommon in diets in developing countries where less meat is eaten and there is high consumption of phytic acid from beans and unleavened whole grain bread.

Foods	Reducing acid present
Cocoa bean and chocolate, spinach, turnip and rhubarb.	Oxalic acid
Whole grains, maize, legumes.	Phytic acid
Tea, beans, cabbage.	Tannins

Nonpolar antioxidants such as eugenol—a major component of oil of cloves—have toxicity limits that can be exceeded with the misuse of undiluted essential oils. Toxicity associated with high doses of water-soluble antioxidants such as ascorbic acid are less of a concern, as these compounds can be excreted rapidly in urine. More seriously, very high doses of some antioxidants may have harmful long-term effects. The beta-carotene and Retinol Efficacy Trial (CARET) study of lung cancer patients found that smokers given supplements containing beta-carotene and vitamin A had increased rates of lung cancer. Subsequent studies confirmed these adverse effects.

These harmful effects may also be seen in non-smokers, as a recent meta-analysis including data from approximately 230,000 patients showed that β-carotene, vitamin A or vitamin E supplementation is associated with increased mortality but saw no significant effect from vitamin C. No health risk was seen when all the randomized controlled studies were examined together, but an increase in mortality was detected when only high-quality and low-bias risk trials were examined separately. As the majority of these low-bias trials dealt with either elderly people, or people with disease, these results may not apply to the general population. This meta-analysis was later repeated and extended by the same authors, with the new analysis published by the Cochrane Collaboration; confirming the previous results. These two publications are consistent with some previous meta-analyzes that also suggested that Vitamin E supplementation increased mortality, and that antioxidant supplements increased the risk of colon cancer. Beta-carotene may also increase lung cancer. Overall, the large number of clinical trials carried out on antioxidant supplements suggest that either these products have no effect on health, or that they cause a small increase in mortality in elderly or vulnerable populations.

While antioxidant supplementation is widely used in attempts to prevent the development of cancer, antioxidants may interfere with cancer treatments, since the environment of cancer cells causes high levels of oxidative stress, making these cells more susceptible to the further oxidative stress induced by treatments. As a result, by reducing the redox stress in cancer cells, antioxidant supplements (and pharmaceuticals) could decrease the effectiveness of radiotherapy and chemotherapy. On the other hand, other reviews have suggested that antioxidants could reduce side effects or increase survival times.

Oxidative Challenge in Biology

A paradox in metabolism is that, while the vast majority of complex life on Earth requires oxygen for its existence, oxygen is a highly reactive molecule that damages living organisms by producing reactive oxygen species. Consequently, organisms contain a complex network of antioxidant metabolites and enzymes that work together to prevent oxidative damage to cellular components such as DNA, proteins and lipids. In general, antioxidant systems either prevent these reactive species from being formed, or remove them before they can damage vital components of the cell. However, reactive oxygen species also have useful cellular functions, such as redox signaling. Thus, the function of antioxidant systems is not to remove oxidants entirely, but instead to keep them at an optimum level.

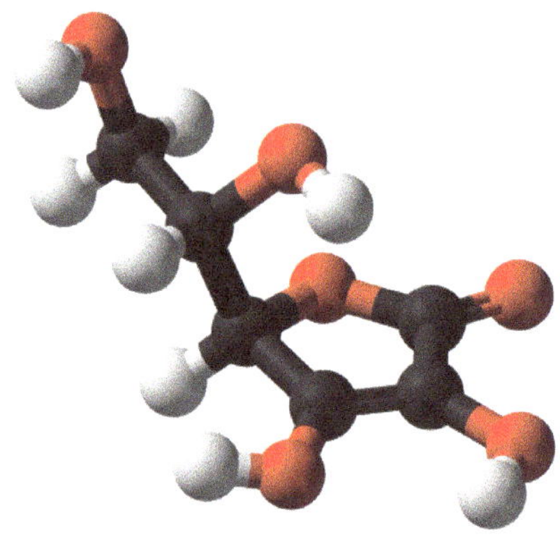

The structure of the antioxidant vitamin ascorbic acid (vitamin C).

The reactive oxygen species produced in cells include hydrogen peroxide (H_2O_2), hypochlorous acid (HClO), and free radicals such as the hydroxyl radical (·OH) and the superoxide anion (O_2^-). The hydroxyl radical is particularly unstable and will react rapidly and non-specifically with most biological molecules. This species is produced from hydrogen peroxide in metal-catalyzed redox reactions such as the Fenton reaction. These oxidants can damage cells by starting chemical chain reactions such as lipid peroxidation, or by oxidizing DNA or proteins. Damage to DNA can cause mutations and possibly cancer, if not reversed by DNA repair mechanisms, while damage to proteins causes enzyme inhibition, denaturation and protein degradation.

The use of oxygen as part of the process for generating metabolic energy produces reactive oxygen species. In this process, the superoxide anion is produced as a by-product of several steps in the electron transport chain. Particularly important is the reduction of coenzyme Q in complex III, since a highly reactive free radical is formed as an intermediate ($Q\cdot^-$). This unstable intermediate can lead to electron "leakage", when electrons jump directly to oxygen and form the superoxide anion, instead of moving through the normal series of well-controlled reactions of the electron transport chain. Peroxide is also produced from the oxidation of reduced flavoproteins, such as complex I. However, although these enzymes can produce oxidants, the relative importance of the electron transfer chain to other processes that generate peroxide is unclear. In plants, algae, and cyanobacteria, reactive oxygen species are also produced during photosynthesis, particularly under conditions of high light intensity. This effect is partly offset by the involvement of carotenoids in photoinhibition, and in algae and cyanobacteria, by large amount of iodide and selenium, which involves these antioxidants reacting with over-reduced forms of the photosynthetic reaction centres to prevent the production of reactive oxygen species.

Metabolites

Antioxidants are classified into two broad divisions, depending on whether they are soluble in water (hydrophilic) or in lipids (lipophilic). In general, water-soluble antioxidants react with oxidants in the cell cytosol and the blood plasma, while lipid-soluble antioxidants protect cell membranes from lipid peroxidation. These compounds may be synthesized in the body or obtained from the diet. The different antioxidants are present at a wide range of concentrations in body fluids and tissues, with some such as glutathione or ubiquinone mostly present within cells, while others such as uric acid are more evenly distributed (see table below). Some antioxidants are only found in a few organisms and these compounds can be important in pathogens and can be virulence factors.

The relative importance and interactions between these different antioxidants is a very complex question, with the various metabolites and enzyme systems having synergistic and interdependent effects on one another. The action of one antioxidant may therefore depend on the proper function of other members of the antioxidant system. The amount of protection provided by any one antioxidant will also depend on its concentration, its reactivity towards the particular reactive oxygen species being considered, and the status of the antioxidants with which it interacts.

Some compounds contribute to antioxidant defense by chelating transition metals and preventing them from catalyzing the production of free radicals in the cell. Particularly important is the ability to sequester iron, which is the function of iron-binding proteins such as transferrin and ferritin. Selenium and zinc are commonly referred to as *antioxidant nutrients*, but these chemical elements have no antioxidant action themselves and are instead required for the activity of some antioxidant enzymes, as is discussed below.

Antioxidant metabolite	Solubility	Concentration in human serum (μM)	Concentration in liver tissue (μmol/kg)
Ascorbic acid (vitamin C)	Water	50 – 60	260 (human)
Glutathione	Water	4	6,400 (human)
Lipoic acid	Water	0.1 – 0.7	4 – 5 (rat)
Uric acid	Water	200 – 400	1,600 (human)
Carotenes	Lipid	β-carotene: 0.5 – 1 retinol (vitamin A): 1 – 3	5 (human, total carotenoids)
α-Tocopherol (vitamin E)	Lipid	10 – 40	50 (human)
Ubiquinol (coenzyme Q)	Lipid	5	200 (human)

Uric Acid

Uric acid is by far the highest concentration antioxidant in human blood. Uric acid (UA) is an antioxidant oxypurine produced from xanthine by the enzyme xanthine oxidase, and is an intermediate product of purine metabolism. In almost all land animals, urate oxidase further catalyzes the oxidation of uric acid to allantoin, but in humans and most higher primates, the urate oxidase gene is nonfunctional, so that UA is not further broken down. The evolutionary reasons for this loss of urate conversion to allantoin remain the topic of active speculation. The antioxidant effects of uric acid have led researchers to suggest this mutation was beneficial to early primates and humans. Studies of high altitude acclimatization support the hypothesis that urate acts as an antioxidant by mitigating the oxidative stress caused by high-altitude hypoxia. In animal studies that investigate diseases facilitated by oxidative stress, introduction of UA both prevents the disease or reduces it, leading researchers to propose this is due to UA's antioxidant properties. Studies of UA's antioxidant mechanism support this proposal.

With respect to multiple sclerosis, Gwen Scott explains the significance of uric acid as an antioxidant by proposing that "Serum UA levels are inversely associated with the incidence of MS in humans because MS patients have low serum UA levels and individuals with hyperuricemia (gout) rarely develop the disease. Moreover, the administration of UA is therapeutic in experimental allergic encephalomyelitis (EAE), an animal model of MS." In sum, while the mechanism of UA as an antioxidant is well-supported, the claim that its levels affect MS risk is still controversial, and requires more research.

Likewise, UA has the highest concentration of any blood antioxidant and provides over half of the total antioxidant capacity of human serum. Uric acid's antioxidant activities are also complex, given that it does not react with some oxidants, such as superoxide, but does act against peroxynitrite, peroxides, and hypochlorous acid. Concerns over elevated UA's contribution to gout must be considered as one of many risk factors. By itself, UA-related risk of gout at high levels (415–530 μmol/L) is only 0.5% per year with an increase to 4.5% per year at UA supersaturation levels (535+ μmol/L). Many of these aforementioned studies determined UA's antioxidant actions within normal physiological levels, and some found antioxidant activity at levels as high as 285 μmol/L.

Vitamin C

Ascorbic acid or "vitamin C" is a monosaccharide oxidation-reduction (redox) catalyst found in both animals and plants. As one of the enzymes needed to make ascorbic acid has been lost by mutation during primate evolution, humans must obtain it from the diet; it is therefore a vitamin. Most other animals are able to produce this compound in their bodies and do not require it in their diets. Ascorbic acid is required for the conversion of the procollagen to collagen by oxidizing proline residues to hydroxyproline. In other cells, it is maintained in its reduced form by reaction with glutathione, which can be catalysed by protein disulfide isomerase and glutaredoxins. Ascorbic acid is a redox catalyst which can reduce, and thereby neutralize, reactive oxygen species such as hydrogen peroxide. In addition to its direct antioxidant effects, ascorbic acid is also a substrate for the redox enzyme ascorbate peroxidase, a function that is particularly important in stress resistance in plants. Ascorbic acid is present at high levels in all parts of plants and can reach concentrations of 20 millimolar in chloroplasts.

Glutathione

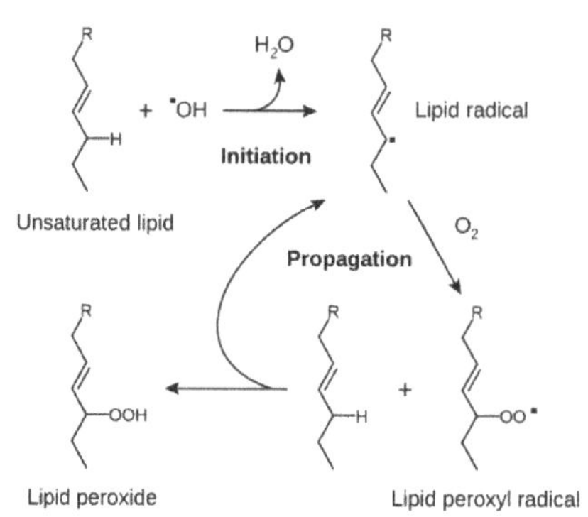

The free radical mechanism of lipid peroxidation.

Glutathione is a cysteine-containing peptide found in most forms of aerobic life. It is not required in the diet and is instead synthesized in cells from its constituent amino acids. Glutathione has antioxidant properties since the thiol group in its cysteine moiety is a reducing agent and can be reversibly oxidized and reduced. In cells, glutathione is maintained in the reduced form by the enzyme glutathione reductase and in turn reduces other metabolites and enzyme systems, such as ascorbate in the glutathione-ascorbate cycle, glutathione peroxidases and glutaredoxins, as well as reacting directly with oxidants. Due to its high concentration and its central role in maintaining the cell's redox state, glutathione is one of the most important cellular antioxidants. In some organisms glutathione is replaced by other thiols, such as by mycothiol in the Actinomycetes, bacillithiol in some Gram-positive bacteria, or by trypanothione in the Kinetoplastids.

Melatonin

Melatonin is a powerful antioxidant. Melatonin easily crosses cell membranes and the blood–brain barrier. Unlike other antioxidants, melatonin does not undergo redox cycling, which is the ability of a molecule to undergo repeated reduction and oxidation. Redox cycling may allow other antioxidants (such as vitamin C) to act as pro-oxidants and promote free radical formation. Melatonin, once oxidized, cannot be reduced to its former state because it forms several stable end-products upon reacting with free radicals. Therefore, it has been referred to as a terminal (or suicidal) antioxidant.

Vitamin E

Vitamin E is the collective name for a set of eight related tocopherols and tocotrienols, which are fat-soluble vitamins with antioxidant properties. Of these, α-tocopherol has been most studied as it has the highest bioavailability, with the body preferentially absorbing and metabolising this form.

It has been claimed that the α-tocopherol form is the most important lipid-soluble antioxidant, and that it protects membranes from oxidation by reacting with lipid radicals produced in the lipid peroxidation chain reaction. This removes the free radical intermediates and prevents the propagation reaction from continuing. This reaction produces oxidised α-tocopheroxyl radicals that can be recycled back to the active reduced form through reduction by other antioxidants, such as ascorbate, retinol or ubiquinol. This is in line with findings showing that α-tocopherol, but not water-soluble antioxidants, efficiently protects glutathione peroxidase 4 (GPX4)-deficient cells from cell death. GPx4 is the only known enzyme that efficiently reduces lipid-hydroperoxides within biological membranes.

However, the roles and importance of the various forms of vitamin E are presently unclear, and it has even been suggested that the most important function of α-tocopherol is as a signaling molecule, with this molecule having no significant role in antioxidant metabolism. The functions of the other forms of vitamin E are even less well-understood, although γ-tocopherol is a nucleophile that may react with electrophilic mutagens, and tocotrienols may be important in protecting neurons from damage.

Pro-oxidant Activities

Antioxidants that are reducing agents can also act as pro-oxidants. For example, vitamin C has antioxidant activity when it reduces oxidizing substances such as hydrogen peroxide, however, it will also reduce metal ions that generate free radicals through the Fenton reaction.

$$2\ Fe^{3+} + \text{Ascorbate} \rightarrow 2\ Fe^{2+} + \text{Dehydroascorbate}$$

$$2\ Fe^{2+} + 2\ H_2O_2 \rightarrow 2\ Fe^{3+} + 2\ OH\cdot + 2\ OH^-$$

The relative importance of the antioxidant and pro-oxidant activities of antioxidants is an area of current research, but vitamin C, which exerts its effects as a vitamin by oxidizing polypeptides, appears to have a mostly antioxidant action in the human body. However, less data is available for other dietary antioxidants, such as vitamin E, or the polyphenols. Likewise, the pathogenesis of diseases involving hyperuricemia likely involve uric acid's direct and indirect pro-oxidant properties.

That is, paradoxically, agents which are normally considered antioxidants can act as conditional pro-oxidants and actually increase oxidative stress. Besides ascorbate, medically important conditional pro-oxidants include uric acid and sulfhydryl amino acids such as homocysteine. Typically, this involves some transition-series metal such as copper or iron as catalyst. The potential role of the pro-oxidant role of uric acid in (e.g.) atherosclerosis and ischemic stroke is considered above. Another example is the postulated role of homocysteine in atherosclerosis.

Negative Health Effects

Some antioxidant supplements may promote disease and increase mortality in humans under certain conditions. Hypothetically, free radicals induce an endogenous response that protects against exogenous radicals (and possibly other toxic compounds). Free radicals may increase life span. This increase may be prevented by antioxidants, providing direct evidence that toxic radicals may mitohormetically exert life extending and health promoting effects.

Enzyme Systems

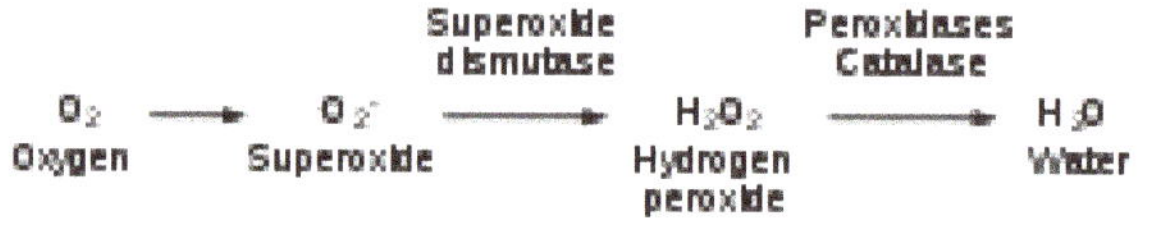

Enzymatic pathway for detoxification of reactive oxygen species.

As with the chemical antioxidants, cells are protected against oxidative stress by an interacting network of antioxidant enzymes. Here, the superoxide released by processes such as oxidative phosphorylation is first converted to hydrogen peroxide and then further reduced to give water. This detoxification pathway is the result of multiple enzymes, with superoxide dismutases catalysing the first step and then catalases and various peroxidases removing hydrogen peroxide. As with antioxidant metabolites, the contributions of these enzymes to antioxidant defenses can be hard to separate from one another, but the generation of transgenic mice lacking just one antioxidant enzyme can be informative.

Superoxide Dismutase, Catalase and Peroxiredoxins

Superoxide dismutases (SODs) are a class of closely related enzymes that catalyze the breakdown of the superoxide anion into oxygen and hydrogen peroxide. SOD enzymes are present in almost all aerobic cells and in extracellular fluids. Superoxide dismutase enzymes contain metal ion co-factors that, depending on the isozyme, can be copper, zinc, manganese or iron. In humans, the copper/zinc SOD is present in the cytosol, while manganese SOD is present in the mitochondrion. There also exists a third form of SOD in extracellular fluids, which contains copper and zinc in its active sites. The mitochondrial isozyme seems to be the most biologically important of these three, since mice lacking this enzyme die soon after birth. In contrast, the mice lacking copper/zinc SOD (Sod1) are viable but have numerous pathologies and a reduced lifespan, while mice without the extracellular SOD have minimal defects (sensitive to hyperoxia). In plants, SOD isozymes are present in the cytosol and mitochondria, with an iron SOD found in chloroplasts that is absent from vertebrates and yeast.

Catalases are enzymes that catalyse the conversion of hydrogen peroxide to water and oxygen, using either an iron or manganese cofactor. This protein is localized to peroxisomes in most eukaryotic cells. Catalase is an unusual enzyme since, although hydrogen peroxide is its only substrate, it follows a ping-pong mechanism. Here, its cofactor is oxidised by one molecule of hydrogen peroxide and then regenerated by transferring the bound oxygen to a second molecule of substrate. Despite its apparent importance in hydrogen peroxide removal, humans with genetic deficiency of catalase — "acatalasemia" — or mice genetically engineered to lack catalase completely, suffer few ill effects.

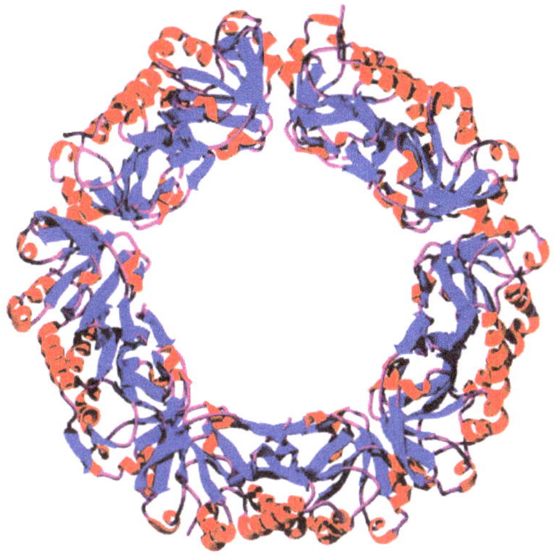

Decameric structure of AhpC, a bacterial 2-cysteine peroxiredoxin from *Salmonella typhimurium*.

Peroxiredoxins are peroxidases that catalyze the reduction of hydrogen peroxide, organic hydroperoxides, as well as peroxynitrite. They are divided into three classes: typical 2-cysteine peroxiredoxins; atypical 2-cysteine peroxiredoxins; and 1-cysteine peroxiredoxins. These enzymes share the same basic catalytic mechanism, in which a redox-active cysteine (the peroxidatic cysteine) in the active site is oxidized to a sulfenic acid by the peroxide substrate. Over-oxidation of this cysteine residue in peroxiredoxins inactivates these enzymes, but this can be reversed by the action of sulfiredoxin. Peroxiredoxins seem to be important in antioxidant metabolism, as mice lacking peroxiredoxin 1 or 2 have shortened lifespan and suffer from hemolytic anaemia, while plants use peroxiredoxins to remove hydrogen peroxide generated in chloroplasts.

Thioredoxin and Glutathione Systems

The thioredoxin system contains the 12-kDa protein thioredoxin and its companion thioredoxin reductase. Proteins related to thioredoxin are present in all sequenced organisms. Plants, such as *Arabidopsis thaliana,* have a particularly great diversity of isoforms. The active site of thioredoxin consists of two neighboring cysteines, as part of a highly conserved CXXC motif, that can cycle between an active dithiol form (reduced) and an oxidized disulfide form. In its active state, thioredoxin acts as an efficient reducing agent, scavenging reactive oxygen species and maintaining other proteins in their reduced state. After being oxidized, the active thioredoxin is regenerated by the action of thioredoxin reductase, using NADPH as an electron donor.

The glutathione system includes glutathione, glutathione reductase, glutathione peroxidases and glutathione *S*-transferases. This system is found in animals, plants and microorganisms. Glutathione peroxidase is an enzyme containing four selenium-cofactors that catalyzes the breakdown of hydrogen peroxide and organic hydroperoxides. There are at least four different glutathione peroxidase isozymes in animals. Glutathione peroxidase 1 is the most abundant and is a very efficient scavenger of hydrogen peroxide, while glutathione peroxidase 4 is most active with lipid hydroperoxides. Surprisingly, glutathione peroxidase 1 is dispensable, as mice lacking this enzyme have normal lifespans, but they are hypersensitive to induced oxidative stress. In addition, the glutathione *S*-transferases show high activity with lipid peroxides. These enzymes are at particularly high levels in the liver and also serve in detoxification metabolism.

Oxidative Stress in Disease

Oxidative stress is thought to contribute to the development of a wide range of diseases including Alzheimer's disease, Parkinson's disease, the pathologies caused by diabetes, rheumatoid arthritis, and neurodegeneration in motor neuron diseases. In many of these cases, it is unclear if oxidants trigger the disease, or if they are produced as a secondary consequence of the disease and from general tissue damage; One case in which this link is particularly well-understood is the role of oxidative stress in cardiovascular disease. Here, low density lipoprotein (LDL) oxidation appears to trigger the process of atherogenesis, which results in atherosclerosis, and finally cardiovascular disease.

Oxidative damage in DNA can cause cancer. Several antioxidant enzymes such as superoxide dismutase, catalase, glutathione peroxidase, glutathione reductase, glutathione S-transferase etc. protect DNA from oxidative stress. It has been proposed that polymorphisms in these enzymes are associated with DNA damage and subsequently the individual's risk of cancer susceptibility.

A low calorie diet extends median and maximum lifespan in many animals. This effect may involve a reduction in oxidative stress. While there is some evidence to support the role of oxidative stress in aging in model organisms such as *Drosophila melanogaster* and *Caenorhabditis elegans*, the evidence in mammals is less clear. Indeed, a 2009 review of experiments in mice concluded that almost all manipulations of antioxidant systems had no effect on aging.

Diets high in fruit and vegetables, and so possibly being rich in antioxidant vitamins, have no established effect on status of health or aging, yet may have more subtle physiological effects, such as modifying cell-to-cell communication.

Uses in Technology

Food Preservatives

Antioxidants are used as food additives to help guard against food deterioration. Exposure to oxygen and sunlight are the two main factors in the oxidation of food, so food is preserved by keeping in the dark and sealing it in containers or even coating it in wax, as with cucumbers. However, as oxygen is also important for plant respiration, storing plant materials in anaerobic conditions produces unpleasant flavors and unappealing colors. Consequently, packaging of fresh fruits and vegetables contains an ~8% oxygen atmosphere. Antioxidants are an especially important class of preservatives as, unlike bacterial or fungal spoilage, oxidation reactions still occur relatively rapidly in frozen or refrigerated food. These preservatives include natural antioxidants such as ascorbic acid (AA, E300) and tocopherols (E306), as well as synthetic antioxidants such as propyl gallate (PG, E310), tertiary butylhydroquinone (TBHQ), butylated hydroxyanisole (BHA, E320) and butylated hydroxytoluene (BHT, E321).

The most common molecules attacked by oxidation are unsaturated fats; oxidation causes them to turn rancid. Since oxidized lipids are often discolored and usually have unpleasant tastes such as metallic or sulfurous flavors, it is important to avoid oxidation in fat-rich foods. Thus, these foods are rarely preserved by drying; instead, they are preserved by smoking, salting or fermenting. Even less fatty foods such as fruits are sprayed with sulfurous antioxidants prior to air drying. Oxidation is often catalyzed by metals, which is why fats such as butter should never be wrapped in aluminium foil or kept in metal containers. Some fatty foods such as olive oil are partially protected from oxidation by their natural content of antioxidants, but remain sensitive to photooxidation. Antioxidant preservatives are also added to fat based cosmetics such as lipstick and moisturizers to prevent rancidity.

Industrial Uses

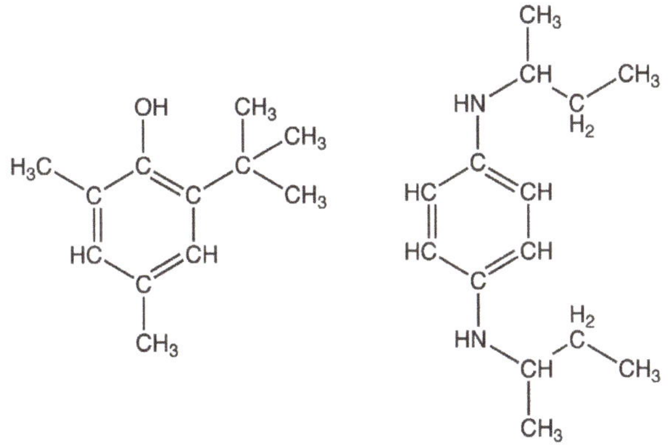

Substituted phenols and derivatives of phenylenediamine are common antioxidants used to inhibit gum formation in gasoline (petrol).

Antioxidants are frequently added to industrial products. A common use is as stabilizers in fuels and lubricants to prevent oxidation, and in gasolines to prevent the polymerization that leads to the formation of engine-fouling residues. In 2007, the worldwide market for industrial antioxidants had a total volume of around 0.88 million tons. This created a revenue of circa 3.7 billion US-dollars (2.4 billion Euros).

They are widely used to prevent the oxidative degradation of polymers such as rubbers, plastics and adhesives that causes a loss of strength and flexibility in these materials. Polymers containing double bonds in their main chains, such as natural rubber and polybutadiene, are especially susceptible to oxidation and ozonolysis. They can be protected by antiozonants. Solid polymer products start to crack on exposed surfaces as the material degrades and the chains break. The mode of cracking varies between oxygen and ozone attack, the former causing a "crazy paving" effect, while ozone attack produces deeper cracks aligned at right angles to the tensile strain in the product. Oxidation and UV degradation are also frequently linked, mainly because UV radiation creates free radicals by bond breakage. The free radicals then react with oxygen to produce peroxy radicals which cause yet further damage, often in a chain reaction. Other polymers susceptible to oxidation include polypropylene and polyethylene. The former is more sensitive owing to the presence of secondary carbon atoms present in every repeat unit. Attack occurs at this point because the free radical formed is more stable than one formed on a primary carbon atom. Oxidation of polyethylene tends to occur at weak links in the chain, such as branch points in low-density polyethylene.

Fuel additive	Components	Applications
AO-22	N,N'-di-2-butyl-1,4-phenylenediamine	Turbine oils, transformer oils, hydraulic fluids, waxes, and greases
AO-24	N,N'-di-2-butyl-1,4-phenylenediamine	Low-temperature oils
AO-29	2,6-di-tert-butyl-4-methylphenol	Turbine oils, transformer oils, hydraulic fluids, waxes, greases, and gasolines
AO-30	2,4-dimethyl-6-tert-butylphenol	Jet fuels and gasolines, including aviation gasolines
AO-31	2,4-dimethyl-6-tert-butylphenol	Jet fuels and gasolines, including aviation gasolines
AO-32	2,4-dimethyl-6-tert-butylphenol and 2,6-di-tert-butyl-4-methylphenol	Jet fuels and gasolines, including aviation gasolines
AO-37	2,6-di-tert-butylphenol	Jet fuels and gasolines, widely approved for aviation fuels

Measurement and Levels in Food

Fruits and vegetables are good sources of antioxidant vitamins.

Antioxidant vitamins are found in vegetables, fruits, eggs, legumes and nuts. Vitamins A, C, and E can be destroyed by long-term storage or prolonged cooking. The effects of cooking and food processing are complex, as these processes can also increase the bioavailability of antioxidants, such as some carotenoids in vegetables. Processed food contains fewer antioxidant vitamins than fresh and uncooked foods, as preparation exposes food to heat and oxygen.

Antioxidant vitamins	Foods containing high levels of antioxidant vitamins
Vitamin C (ascorbic acid)	Fresh or frozen fruits and vegetables
Vitamin E (tocopherols, tocotrienols)	Vegetable oils, nuts and seeds
Carotenoids (carotenes as provitamin A)	Fruit, vegetables and eggs

Other antioxidants are not vitamins and are instead made in the body. For example, ubiquinol (coenzyme Q) is poorly absorbed from the gut and is made in humans through the mevalonate pathway. Another example is glutathione, which is made from amino acids. As any glutathione in the gut is broken down to free cysteine, glycine and glutamic acid before being absorbed, even large oral doses have little effect on the concentration of glutathione in the body. Although large amounts of sulfur-containing amino acids such as acetylcysteine can increase glutathione, no evidence exists that eating high levels of these glutathione precursors is beneficial for healthy adults. Supplying more of these precursors may be useful as part of the treatment of some diseases, such as acute respiratory distress syndrome, protein-energy malnutrition, or preventing the liver damage produced by paracetamol overdose.

Other compounds in the diet can alter the levels of antioxidants by acting as pro-oxidants whereby consuming the compound may cause oxidative stress, possibly resulting in higher levels of antioxidant enzymes.

Invalidation of ORAC

Measurement of antioxidant content in food is not a straightforward process, as this is a diverse group of compounds with different reactivities to various reactive oxygen species. In food science, the oxygen radical absorbance capacity (ORAC) used to be the industry standard for antioxidant strength of whole foods, juices and food additives. However, the United States Department of Agriculture (USDA) withdrew these ratings in 2012 as biologically invalid, stating that no physiological proof *in vivo* existed to support the free-radical theory, especially for polyphenols. Consequently, the ORAC method, derived only from *in vitro* experiments, is no longer considered relevant to human diets or biology.

Alternative *in vitro* measurements include the Folin-Ciocalteu reagent, and the Trolox equivalent antioxidant capacity assay.

History

As part of their adaptation from marine life, terrestrial plants began producing non-marine antioxidants such as ascorbic acid (vitamin C), polyphenols and tocopherols. The evolution of angiosperm plants between 50 and 200 million years ago resulted in the development of many antioxidant pigments – particularly during the Jurassic period – as chemical defences against reactive oxygen species that are byproducts of photosynthesis. Originally, the term antioxidant specifically referred to a chemical that prevented the consumption of oxygen. In the late 19th and early 20th centuries, extensive study concentrated on the use of antioxidants in important industrial processes, such as the prevention of metal corrosion, the vulcanization of rubber, and the polymerization of fuels in the fouling of internal combustion engines.

Early research on the role of antioxidants in biology focused on their use in preventing the oxidation of unsaturated fats, which is the cause of rancidity. Antioxidant activity could be measured simply by placing the fat in a closed container with oxygen and measuring the rate of oxygen consumption. However, it was the identification of vitamins A, C, and E as antioxidants that revolutionized the field and led to the realization of the importance of antioxidants in the biochemistry of living organisms. The possible mechanisms of action of antioxidants were first explored when it was recognized that a substance with anti-oxidative activity is likely to be one that is itself readily oxidized. Research into how vitamin E prevents the process of lipid peroxidation led to the identification of antioxidants as reducing agents that prevent oxidative reactions, often by scavenging reactive oxygen species before they can damage cells.

References

- Anderson, Oscar Edward (1953). Refrigeration in America; a history of a new technology and its impact. [Princeton]: Published for the University of Cincinnati by Princeton University Press. pp. 5–6. ISBN 0804616213.
- Freidberg, Susanne (2010). Fresh : a perishable history (1st Harvard University Press pbk. ed.). Cambridge, Mass.: Belknap. pp. 20–23. ISBN 0674057228.
- Arora, Ramesh Chandra. "Mechanical vapour compression refrigeration". Refrigeration and Air Conditioning. New Delhi, India: PHI Learning. p. 3. ISBN 81-203-3915-0.
- Freidberg, Susanne (2010). Fresh : a perishable history (1st Harvard University Press pbk. ed.). Cambridge, Mass.: Belknap. p. 23. ISBN 0674057228.
- Anderson, Oscar Edward (1953). Refrigeration in America; a history of a new technology and its impact. [Princeton]: Published for the University of Cincinnati by Princeton University Press. p. 25. ISBN 0804616213.
- Freidberg, Susanne (2010). Fresh : a perishable history (1st Harvard University Press pbk. ed.). Cambridge, Mass.: Belknap. p. 25. ISBN 0674057228.
- Anderson, Oscar Edward (1953). Refrigeration in America; a history of a new technology and its impact. [Princeton]: Published for the University of Cincinnati by Princeton University Press. p. 156. ISBN 0804616213.
- Rockoff, Gary M. Walton, Hugh (2010). History of the American Economy (11th ed.). Mason, Ohio: South-Western/Cengage Learning. pp. 336–368. ISBN 0324786611.
- Islami, F (2009). "Pickled vegetables and the risk of oesophageal cancer: a meta-analysis". British Journal of Cancer. 101 (9): 1641–1647. doi:10.1038/sj.bjc.6605372. PMC 2778505. PMID 19862003. Retrieved 2013-05-14.
- Polypropylene: the definitive user's guide and databook. Clive Maier, Teresa Calafut, 1998 ISBN 1-884207-58-8, ISBN 978-1-884207-58-7 Length: 432 pages. Books.google.com. Retrieved 2013-07-08.
- "Withdrawn: Oxygen Radical Absorbance Capacity (ORAC) of Selected Foods, Release 2 (2010)". United States Department of Agriculture, Agricultural Research Service. 16 May 2012. Retrieved 13 June 2012.

4

Modern Techniques of Food Safety

Food safety is the basic concern of humans and with evolving times, better techniques have been introduced in this regard. Some of the modern techniques are pasteurization, vacuum packing, food irradiation, bio preservation and acidity regulator. The chapter serves as a source to understand these major categories.

Pasteurization

Pasteurization or pasteurisation is a process that kills microbes (mainly bacteria) in food and drink, such as milk, juice, canned food, and others.

Cream pasteurizing and cooling coils at Murgon Butter Factory, 1939

It was invented by French scientist Louis Pasteur during the nineteenth century. In 1864 Pasteur discovered that heating beer and wine was enough to kill most of the bacteria that caused spoilage, preventing these beverages from turning sour. The process achieves this by eliminating pathogenic microbes and lowering microbial numbers to prolong the quality of the beverage. Today, pasteurisation is used widely in the dairy industry and other food processing industries to achieve food preservation and food safety.

Unlike sterilization, pasteurization is not intended to kill all microorganisms in the food. Instead, it aims to reduce the number of viable pathogens so they are unlikely to cause disease (assuming the pasteurized product is stored as indicated and is consumed before its expiration date). Commercial-scale sterilization of food is not common because it adversely affects the taste and quality of the product. Certain foods, such as dairy products, may be superheated to ensure pathogenic microbes are destroyed.

Pasteurization Conditions

Minimum pasteurization requirements for milk products are shown in Table 1 below, and are based on regulations outlined in the Grade A Pasteurized Milk Ordinance (PMO). These conditions were determined to be the minimum processing conditions needed to kill Coxiella burnetii, the organism that causes Q fever in humans, which is the most heat resistant pathogen currently recognized in milk. Milk can be pasteurized using processing times and temperatures greater than the required minimums.

Pasteurization can be done as a batch or a continuous process. A vat pasteurizer consists of a temperature-controlled, closed vat. The milk is pumped into the vat, the milk is heated to the appropriate temperature and held at that temperature for the appropriate time and then cooled. The cooled milk is then pumped out of the vat to the rest of the processing line, for example to the bottling station or cheese vat. Batch pasteurization is still used in some smaller processing plants. The most common process used for fluid milk is the continuous process. The milk is pumped from the raw milk silo to a holding tank that feeds into the continuous pasteurization system. The milk continuously flows from the tank through a series of thin plates that heat up the milk to the appropriate temperature. The milk flow system is set up to make sure that the milk stays at the pasteurization temperature for the appropriate time before it flows through the cooling area of the pasteurizer. The cooled milk then flows to the rest of the processing line, for example to the bottling station. There are several options for temperatures and times available for continuous processing of refrigerated fluid milk. Although processing conditions are defined for temperatures above 200 °F, they are rarely used because they can impart an undesirable cooked flavor to milk.

Alcoholic Beverages

The process of heating wine for preservation purposes has been known in China since 1117, and was documented in Japan in the diary *Tamonin-nikki*, written by a series of monks between 1478 and 1618.

Much later, in 1768, an Italian priest and scientist Lazzaro Spallanzani proved experimentally that heat killed bacteria, and that they do not re-appear if the product is hermetically sealed. In 1795, a Parisian chef and confectioner named Nicolas Appert began experimenting with ways to preserve foodstuffs, succeeding with soups, vegetables, juices, dairy products, jellies, jams, and syrups. He placed the food in glass jars, sealed them with cork and sealing wax and placed them in boiling water. In that same year, the French military offered a cash prize of 12,000 francs for a new method to preserve food. After some 14 or 15 years of experimenting, Appert submitted his invention and won the prize in January 1810. Later that year, Appert published *L'Art de conserver les substances animales et végétales* (or *The Art of Preserving Animal and Vegetable Substances*). This was the first cookbook of its kind on modern food preservation methods.

La Maison Appert (English: The House of Appert), in the town of Massy, near Paris, became the first food-bottling factory in the world, preserving a variety of food in sealed bottles. Appert's method was to fill thick, large-mouthed glass bottles with produce of every description, ranging from beef and fowl to eggs, milk and prepared dishes. His greatest success for publicity was an entire sheep. He left air space at the top of the bottle, and the cork would then be sealed firmly in the jar by using a vise. The bottle was then wrapped in canvas to protect it, while it was dunked into boiling water and then boiled for as much time as Appert deemed appropriate for cooking the contents thoroughly. Appert patented his method, sometimes called in his honor *appertisation*.

Appert's method was so simple and workable that it quickly became widespread. In 1810, British inventor and merchant Peter Durand, also of French origin, patented his own method, but this time in a tin can, so creating the modern-day process of canning foods. In 1812, Englishmen Bryan Donkin and John Hall purchased both patents and began producing preserves. Just a decade later, Appert's method of canning had made its way to America. Tin can production was not common until the beginning of the 20th century, partly because a hammer and chisel were needed to open cans until the invention of a can opener by an inventor named Yates in 1855.

Appert's preservation by boiling involved heating the food to an unnecessarily high temperature, and for an unnecessarily long time, which could destroy some of the flavor of the preserved food.

A less aggressive method was developed by the French chemist Louis Pasteur during an 1864 summer holiday in Arbois. To remedy the frequent acidity of the local wines, he found out experimentally that it is sufficient to heat a young wine to only about 50–60 °C (122–140 °F) for a brief time to kill the microbes, and that the wine could subsequently be aged without sacrificing the final quality. In honour of Pasteur, the process became known as "pasteurization" Pasteurization was originally used as a way of preventing wine and beer from souring, and it would be many years before milk was pasteurized. In the United States in the 1870s, it was common for milk to contain substances intended to mask spoilage before milk was regulated.

Milk

400 lbs of milk in cheese vat

Milk is an excellent medium for microbial growth, and when stored at ambient temperature bacteria and other pathogens soon proliferate.

The US Centers for Disease Control (CDC) says improperly handled raw milk is responsible for nearly three times more hospitalizations than any other food-borne disease source, making it one of the world's most dangerous food products. Diseases prevented by pasteurization can include tuberculosis, brucellosis, diphtheria, scarlet fever, and Q-fever; it also kills the harmful bacteria *Salmonella*, *Listeria*, *Yersinia*, *Campylobacter*, *Staphylococcus aureus*, and *Escherichia coli* O157:H7, among others.

Pasteurization is the reason for milk's extended shelf life. High-temperature, short-time (HTST) pasteurized milk typically has a refrigerated shelf life of two to three weeks, whereas ultra-pasteurized milk can last much longer, sometimes two to three months. When ultra-heat treatment (UHT) is combined with sterile handling and container technology (such as aseptic packaging), it can even be stored unrefrigerated for up to 9 months.

History

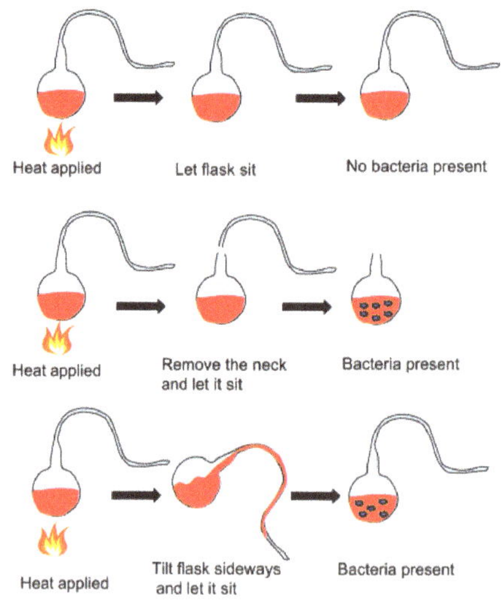

Louis Pasteur's pasteurization experiment illustrates the fact that the spoilage of liquid was caused by particles in the air rather than the air itself. These experiments were important pieces of evidence supporting the idea of Germ Theory of Disease.

Before the widespread urban growth caused by industrialization, people kept dairy cows even in urban areas and the short time period between production and consumption minimized the disease risk of drinking raw milk. As urban densities increased and supply chains lengthened to the distance from country to city, raw milk (often days old) became recognised as a source of disease. For example, between 1912 and 1937 some 65,000 people died of tuberculosis contracted from consuming milk in England and Wales alone.

Developed countries adopted milk pasteurization to prevent such disease and loss of life, and as a result milk is now widely considered one of the safest foods. A traditional form of pasteurization by scalding and straining of cream to increase the keeping qualities of butter was practiced in England before 1773 and was introduced to Boston in the US by 1773, although it was not widely practiced in the United States for the next 20 years. It was still being referred to as a "new" process in American newspapers as late as 1802. Pasteurization of milk was suggested by Franz von Soxhlet in 1886. In the early 20th century, Milton Joseph Rosenau, established the standards (i.e. low temperature, slow heating at 60 °C (140 °F) for 20 minutes) for the pasteurization of milk, while at the United States Marine Hospital Service, notably in his publication of The Milk Question (1912).

States in the U.S.A. began enacting mandatory dairy pasteurization laws with the first in 1947, and in 1973 the U.S. Federal Government required pasteurization of milk used in any interstate commerce.

Process

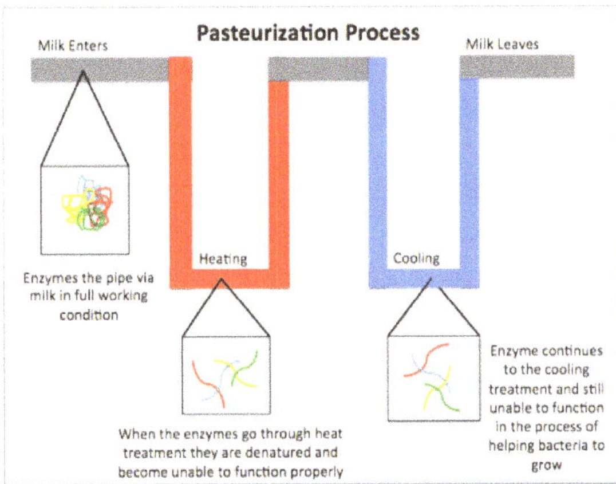

General overview of the pasteurization process. The milk starts at the left and enters the piping with functioning enzymes that, when heat treated, become denatured and stop the enzymes from functioning. This helps to stop pathogen growth by stopping the functionality of the cell. The cooling process helps stop the milk from undergoing the Maillard reaction and caramelization. The pasteurization process also has the ability to heat the cells to the point that they burst from pressure build up.

Older pasteurization methods used temperatures below boiling, since at very high temperatures, micelles of the milk protein casein irreversibly aggregate, or *curdle*. Newer methods use higher temperature, but shorten the time. Among the pasteurization methods listed below, the two main types of pasteurization used today are high-temperature, short-time (HTST, also known as "flash") and extended shelf life (ESL):

- HTST milk is forced between metal plates or through pipes heated on the outside by hot water, and the milk is heated to 72 °C (161 °F) for 15 seconds. Milk simply labeled "pasteurized" is usually treated with the HTST method.
- UHT, also known as ultra-heat-treating, processing holds the milk at a temperature of 140 °C (284 °F) for four seconds. During UHT processing milk is sterilized and not pasteurized. This process lets consumers store milk or juice for several months without refrigeration. The process is achieved by spraying the milk or juice through a nozzle into a chamber filled with high-temperature steam under pressure. After the temperature reaches 140 °C the fluid is cooled instantly in a vacuum chamber, and packed in a pre-sterilized airtight container. Milk labeled "ultra-pasteurized" or simply "UHT" has been treated with the UHT method.
- ESL milk has a microbial filtration step and lower temperatures than UHT milk. Since 2007, it is no longer a legal requirement in European countries (for example in Germany) to declare ESL milk as ultra-heated; consequently, it is now often labeled as "fresh milk"

and just advertised as having an "extended shelf life," making it increasingly difficult to distinguish ESL milk from traditionally pasteurized fresh milk.

- A less conventional, but US FDA-legal, alternative (typically for home pasteurization) is to heat milk at 63 °C (145 °F) for 30 minutes.

Pasteurization methods are usually standardized and controlled by national food safety agencies (such as the USDA in the United States and the Food Standards Agency in the United Kingdom). These agencies require that milk be HTST pasteurized to qualify for the pasteurized label. Dairy product standards differ, depending on fat content and intended usage. For example, pasteurization standards for cream differ from standards for fluid milk, and standards for pasteurizing cheese are designed to preserve the enzyme phosphatase, which aids cutting. In Canada, all milk produced at a processor and intended for consumption must be pasteurized, which legally requires that it be heated to at least 72 °C for at least 16 seconds, then cooling it to 4 °C to ensure any harmful bacteria are destroyed. The UK Dairy Products Hygiene Regulations 1995 requires that milk be heat treated for 15 seconds at 71.7 °C or other effective time/temperature combination.

Some older references point to one or multiple cycles of heating and cooling (to ambient temperature or below) as either a definition of pasteurisation or a general method thereof.

A process similar to pasteurization is thermization, which uses lower temperatures to kill bacteria in milk. It allows a milk product, such as cheese, to retain more of the original taste, but thermized foods are not considered pasteurized by food regulators.

Microwave Volumetric Heating

Microwave volumetric heating (MVH) is the newest available pasteurization technology. It uses microwaves to heat liquids, suspensions, or semi-solids in a continuous flow. Because MVH delivers energy evenly and deeply into the whole body of a flowing product, it allows for gentler and shorter heating, so that almost all heat-sensitive substances in the milk are preserved.

Efficiency

The HTST pasteurization standard was designed to achieve a five-log reduction, killing 99.999% of the number of viable micro-organisms in milk. This is considered adequate for destroying almost all yeasts, molds, and common spoilage bacteria and also to ensure adequate destruction of common pathogenic, heat-resistant organisms (including *Mycobacterium tuberculosis*, which causes tuberculosis, but not *Coxiella burnetii,* which causes Q fever). As a precaution, modern equipment tests and identifies bacteria in milk being processed. HTST pasteurization processes must be designed so the milk is heated evenly, and no part of the milk is subject to a shorter time or a lower temperature.

Even pasteurization without quality control can be effective, though this is generally not permitted for human consumption; a study of farms feeding calves on pasteurized waste milk using a mixture of pasteurization technologies (none of which were routinely monitored for performance) found the resulting pasteurized milk to meet safety requirements at least 92% of the time.

An effect of the heating of pasteurization is that some vitamin, mineral, and beneficial (or probiotic) bacteria is lost. Soluble calcium and phosphorus levels decrease by 5%, thiamine (vitamin B_1) and vitamin B_{12} (cobalamin) levels by 10%, and vitamin C levels by 20%. These losses are not significant nutritionally.

Verification

Direct microbiological techniques are the ultimate measurement of pathogen contamination, but these are costly and time-consuming (24–48 hours), which means that products are able to spoil by the time pasteurization is verified.

As a result of the unsuitability of microbiological techniques, milk pasteurization efficacy is typically monitored by checking for the presence of alkaline phosphatase, which is denatured by pasteurization. B. tuberculosis, the bacterium that requires the highest temperature to be killed of all milk pathogens is killed at ranges of temperature and time similar to those that denature alkaline phosphatase. For this reason, presence of alkaline phosphatase is an ideal indicator of pasteurization efficacy.

Phosphatase denaturing was originally monitored using a phenol-phosphate substrate. When hydrolysed by the enzyme these compounds liberate phenols, which were then reacted with dibromoquinonechlorimide to give a colour change, which itself was measured by checking absorption at 610 nm (spectrophotometry). Some of the phenols used were inherently coloured (phenolpthalein, nitrophenol) and were simply assayed unreacted. Spectrophotometric analysis is satisfactory but is of relatively low accuracy because many natural products are coloured. For this reason, modern systems (since 1990) use fluorometry which is able to detect much lower levels of raw milk contamination.

Unpasteurized Milk

According to the United States Centers for Disease Control between 1998 and 2011 79% of the dairy related outbreaks were due to raw milk or cheese products. They report 148 outbreaks, 2,384 illnesses (284 requiring hospitalizations) as well as 2 deaths due to raw milk or cheese products during the same time period.

Consumer Acceptance

Although pasteurization has been practiced for a long time, some consumers contend that they should have the right to buy and sell un-pasteurized milk if they want to.

Some consumers also point out that government-enforced pasteurization law has been used as a tool for large business to shut out competition from smaller producers.

Vacuum Packing

Vacuum packing is a method of packaging that removes air from the package prior to sealing. This method involves (manually or automatically) placing items in a plastic film package, removing air from inside, and sealing the package. Shrink film is sometimes used to have a tight fit to the contents. The intent of vacuum packing is usually to remove oxygen from the container to extend the shelf life of foods and, with flexible package forms, to reduce the volume of the contents and package.

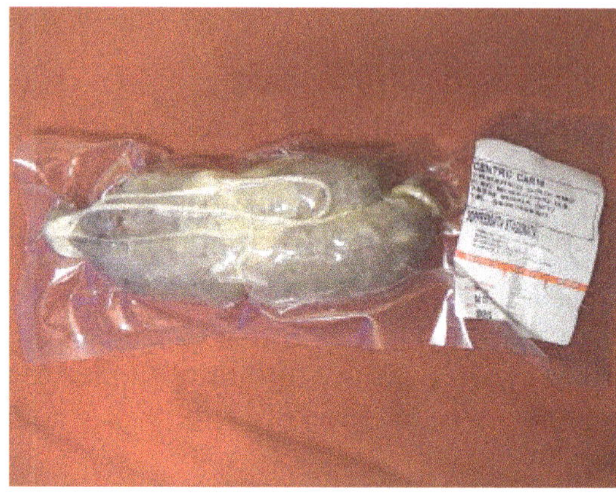

Vacuum packing reduces atmospheric oxygen, limiting the growth of aerobic bacteria or fungi, and preventing the evaporation of volatile components. It is also commonly used to store dry foods over a long period of time, such as cereals, nuts, cured meats, cheese, smoked fish, coffee, and potato chips (crisps). On a more short term basis, vacuum packing can also be used to store fresh foods, such as vegetables, meats, and liquids, because it inhibits bacterial growth.

Vacuum packing greatly reduces the bulk of non-food items. For example, clothing and bedding can be stored in bags evacuated with a domestic vacuum cleaner or a dedicated vacuum sealer. This technique is sometimes used to compact household waste, for example where a charge is made for each full bag collected.

Vacuum packaging products, using plastic bags, canisters, bottles, or mason jars, are available for home use.

For delicate food items which might be crushed by the vacuum packing process (such as potato chips), an alternative is to replace the interior gas with nitrogen. This has the same effect of inhibiting deterioration due to the removal of oxygen.

External Sealers

External vacuum sealers involve a bag being attached to the vacuum-sealing machine externally. The machine will remove the air and seal the bag, which is all done outside the machine. A heat sealer is often used to seal the pack.

Single Vacuum Chamber Machines

Single chamber sealers require the entire product to be placed within the machine. Like external sealers, a plastic bag is typically used for packaging. Once the product is placed in the machine, the lid is closed and air is removed. Then, there is a heat seal inside the chamber that will seal the bag, after sealing the bag the chamber is refilled with air by the automatic opening of a vent to the outside. This oncoming pressure squeezes all remaining air in the bag. The lid is then opened and the product removed. Chamber sealers are typically used for low-to-medium-volume packaging, and also have the capability to vacuum seal liquids.

This video shows vacuum packaging of organic rice.

Tabletop Vacuum Packaging Machine

Double Vacuum Chamber Machines

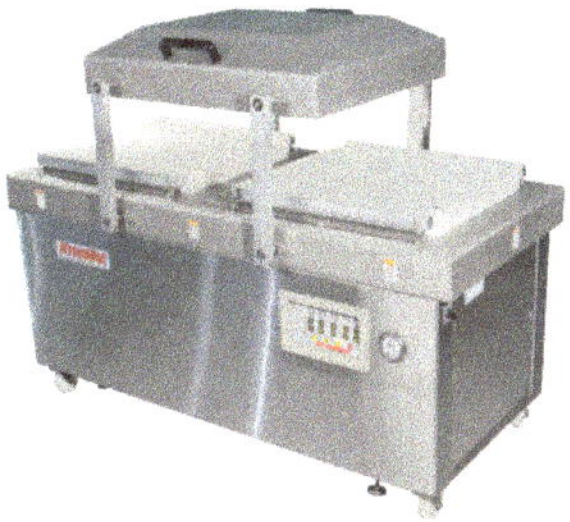

Double Chamber Vacuum Packaging Machine

Double chamber sealers require the entire product to be placed in a plastic bag within the machine. Once the product is placed in the machine on the seal bar, the lid is closed and air is removed. Then a seal bar inside the chamber seals the product in the bag, after sealing the bag the chamber is refilled with air by the automatic opening of a vent to the outside. This oncoming pressure squeezes all remaining air in the bag. The lid is then opened and the product removed. Double chamber sealers are typically used for medium-volume packaging, and also have the capability to vacuum seal liquids. The lid generally swings from one side to another, increasing production speed over a single chamber model. Double chamber vacuum packaging machines generally have either spring-weighted lids or fully automatic lids.

Double chamber vacuum packaging machines are commonly used for:

- Fresh Meat

- Processed Meat
- Cheese (hard and soft)
- Candy & Chocolate
- Empty Cans (it's a example of atmospheric pressure)

Automatic Belt Vacuum Chamber Machines

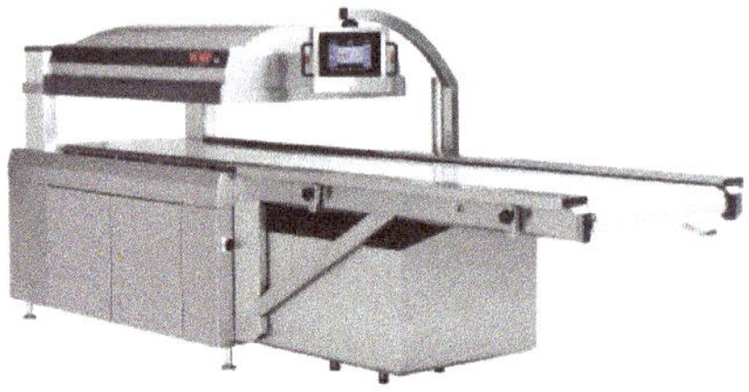

Automatic Belt Vacuum Chamber Machine. Automatic belt vacuum chamber machines offer vastly increased speed and automation and accommodate large products.

Automatic belt chamber sealers require the entire product to be placed in a plastic bag or flow wrapped pouch within the machine. The product travels on the conveyor belt, it is automatically positioned in the machine on the seal bar, the lid is closed and air is removed. Then a seal bar inside the chamber seals the product in the bag. After sealing the bag, the chamber is refilled with air by the automatic opening of a vent to the outside. This oncoming pressure squeezes all remaining air in the bag. The lid is then opened and the product removed. Automatic belt vacuum chamber machines are typically used for high-speed packaging of large items, and also have the capability to vacuum seal liquids. The lid generally travels straight up and down.

Automatic belt vacuum chamber packaging machines are commonly used for:

- Fresh Meat (large portions)
- Processed Meat
- Large Sausage logs
- Cheese (hard and soft)

Thermoforming (Rollstock) Vacuum Packaging Machines

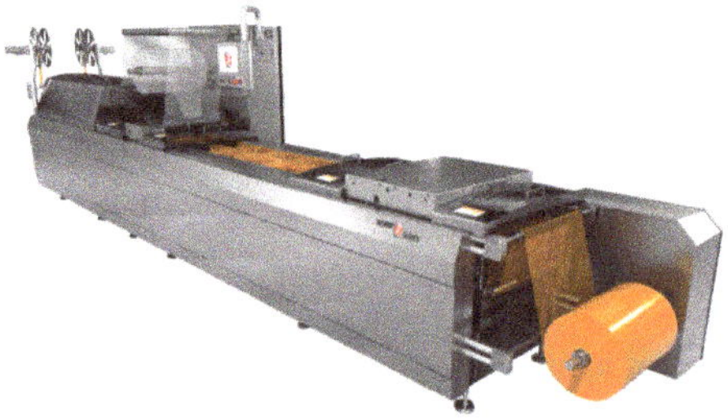

Thermoform packaging machines are used in larger production facilities for vacuum packaging products.

Vacuum Packaging in large production facilities can be done with thermoforming machines. These are Form-Fill-Seal style machines that form the package from rolls of packaging film (webbing). Products are loaded into the thermoformed pockets, the top web is laid and sealed under a vacuum, producing vacuum packaged products. Thermoforming can greatly increase packaging production speed. Thermoformed plastics can be customized for size, color, clarity, and shape to fit products perfectly, creating a consistent appearance. Some common uses for Thermoforming in vacuum packaging include:

- Fresh & Marinated Meat
- Sausage
- Cheese
- Candy / Chocolate
- Grain
- Grab-and-Go Snacks (beef jerky, snack sticks)
- Pharmaceutical and Medical Products
- Coins / Collectables

Shelf Life

Depending on the product, the shelf life of vacuum packaged products can exceed normal bagged or wrapped packages. Beef can last up to six weeks refrigerated, and much longer when frozen.

High Barrier Shrink Vacuum Bags

The amount of shelf life enhanced by a vacuum bag is dependent on the structure in the material. A standard vacuum bag is composed of a PA/PE structure where PA is for puncture resistance and PE is for sealing. The high barrier category includes the usage of more layers focused on the prevention of oxygen permeability, and therefore shelf life protection. There are two materials used in high barrier structures, polyvinylidene chloride (PVDC) and ethylene vinyl alcohol (EVOH). Shelf life indication can be effectively measured by how many cubic centimeters of oxygen can permeate through 1 square meter of material over a 24-hour period. A standard PA/PE bag allows on average 100 cubic centimeters, PVDC allows on average over 10, and EVOH on average 1 cubic centimeter. Multi-layer structures allow the ability to use strong oxygen-barrier materials for enhanced shelf life protection. The PremiumPack structure is a good example of EVOH based high barrier shrink material.

Preventing Freezer Burn

When foods are frozen without preparation, freezer burn can occur. It happens when the surface of the food is dehydrated, and this leads to a dried and leathery appearance. Freezer burn also ruins the flavor and texture of foods. Vacuum packing reduces freezer burn by preventing the food from exposure to the cold, dry air.

Sous-vide Cooking

Vacuum packaging also allows for a special cooking method, sous-vide. Sous-vide, French for *under vacuum*, involves poaching food that is vacuum sealed in a plastic bag.

Food Safety

In an oxygen-depleted environment, anaerobic bacteria can proliferate, potentially causing food-safety issues. Vacuum packing is often used in combination with other packaging and food processing techniques.

Preservative

A preservative is a substance or a chemical that is added to products such as food, beverages, pharmaceutical drugs, paints, biological samples, cosmetics, wood, and many other products to prevent decomposition by microbial growth or by undesirable chemical changes. In general, preservation is implemented in two modes, chemical and physical. Chemical preservation entails adding chemical compounds to the product.Physical preservation entails processes such as refrigeration or drying. Preservative food additives reduce the risk of foodborne infections, decrease microbial spoilage, and preserve fresh attributes and nutritional quality. Some physical techniques for food preservation include dehydration, UV-C radiation, freeze-drying, and refrigeration. Chemical preservation and physical preservation techniques are sometimes combined.

Antimicrobial Additives

Antimicrobial preservatives prevent degradation by bacteria. This method is the most traditional and ancient type of preserving—ancient methods such as pickling and adding honey prevent microorganism growth by modifying the pH level. The most commonly used antimicrobial preservative is lactic acid. Common antimicrobial preservatives are presented in the table. Nitrates and nitrites are also antimicrobial. The detailed mechanism of these chemical compounds range from inhibiting growth of the bacteria to the inhibition of specific enzymes.

E number	chemical compound	comment
E201 – E203	benzoic acid, sodium benzoate	used in acidic foods such as jams, salad dressing, juices, pickles, carbonated drinks, soy sauce
E214 – E219	hydroxybenzoate and derivatives	stable at a broad pH range
E270	lactic acid	-
E249 – E250	nitrite	used in meats to prevent botulism toxin
E251 – E252	nitrate	used in meats
E280 – E283	propionic acid and sodium propionate	baked goods
E220 – E227	sulfur dioxide and sulfites	common for fruits
E200 – E203	sorbic acid and sodium sorbate	common for cheese, wine, baked goods

Antioxidants

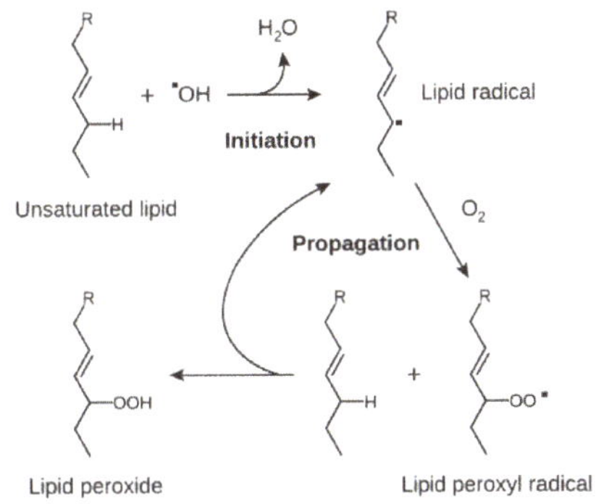

The free radical pathway for the first phase of the oxidative rancidification of fats. This process is slowed by antioxidants.

The oxidation process spoils most food, especially those with a high fat content. Fats quickly turn rancid when exposed to oxygen. Antioxidants prevent or inhibit the oxidation process. The most common antioxidant additives are ascorbic acid (vitamin C) and ascorbates. Thus, antioxidants are commonly added to oils, cheese, and chips. Other antioxidants include the phenol derivatives BHA, BHT, TBHQ and propyl gallate. These agents suppress the formation of hydroperoxides. Other preservatives include ethanol and methylchloroisothiazolinone.

E number	chemical compound	comment
E300-304	ascorbic acid, sodium ascorbate	cheese, chips
E321	butylated hydroxytoluene, butylated hydroxyanisole	also used in food packaging
E310-312	gallic acid and sodium gallate	oxygen scavenger
E220 – E227	sulfur dioxide and sulfites	beverages, wine
E306 – E309	tocopherols	vitamin E activity

A variety of agents are added to sequester (deactivate) metal ions that otherwise catalyze the oxidation of fats. Common sequestering agents are disodium EDTA, citric acid (and citrates), tartaric acid, and lecithin.

Nonsynthetic Compounds for Food Preservation

Citric and ascorbic acids target enzymes that degrade fruits and vegetables, e.g., phenolase which turns surfaces of cut apples and potatoes brown. Ascorbic acid and tocopherol, which are vitamins, are common preservatives. Smoking entails exposing food to a variety of phenols, which are antioxidants. Natural preservatives include rosemary extract, hops, salt, sugar, vinegar, alcohol, diatomaceous earth and castor oil.

Traditional preservatives, such as sodium benzoate have raised health concerns in the past. Benzoate was shown in a study to cause hypersensitivity in some asthma sufferers. This has caused reexamination of natural preservatives which occur in vegetables.

History

Preservatives have been used since prehistoric times. Smoked meat for example has phenols and other chemicals that retard spoilage. The preservation of foods has evolved greatly over the centuries and has been instrumental in increasing food security. The use of preservatives other than traditional oils, salts, paints, etc. in food began in the late 19th century, but was not widespread until the 20th century.

The use of food preservatives varies greatly depending on the country. Many developing countries that do not have strong governments to regulate food additives face either harmful levels of preservatives in foods or a complete avoidance of foods that are considered unnatural or foreign. These countries have also proven useful in case studies surrounding chemical preservatives, as they have been only recently introduced. In urban slums of highly populated countries, the knowledge about contents of food tends to be extremely low, despite consumption of these imported foods.

Public Awareness of Food Preservation

Public awareness of food preservatives is uneven. Americans have a perception that food-borne illnesses happen more often in other countries. This may be true, but the occurrence of illnesses, hospitalizations, and deaths are still high. It is estimated by the Center for Disease Control (CDC) that each year there are 76 million illnesses, 325,000 hospitalizations, and 5,000 deaths linked to food-borne illness.

The increasing demand for ready-to-eat fresh food products has led to challenges for food distributors regarding the safety and quality of their foods. Artificial preservatives meet some of these challenges by preserving freshness for longer periods of time, but these preservatives can cause negative side-effects as well. Sodium nitrite is a preservative used in lunch meats, hams, sausages, hot dogs, and bacon to prevent botulism. It serves the important function of controlling the bacteria that cause botulism, but sodium nitrite can react with proteins, or during cooking at high heats, to form carcinogenic N-nitrosamines. It has also been linked to cancer in lab animals. The commonly used sodium benzoate has been found to extend the shelf life of bottled tomato paste to 40 weeks without loss of quality. However, it can form the carcinogen benzene when combined with vitamin C. Many food manufacturers have reformed their products to eliminate this combination, but a risk still exists. Consumption of sodium benzoate may also cause hyperactivity. For over 30 years, there has been a debate about whether or not preservatives and other food additives can cause hyperactivity. Studies have found that there may be increases in hyperactivity amongst children who consume artificial colorings and benzoate preservatives and who are already genetically predisposed to hyperactivity, but these studies were not entirely conclusive. Hyperactivity only increased moderately, and it was not determined if the preservatives, colorings, or a combination of the two were responsible for the increase.

Food Irradiation

Food irradiation is the process of exposing foodstuffs to ionizing radiation. Ionizing radiation is energy that can be transmitted without direct contact (radiation) capable of freeing electrons from their atomic bonds (ionization) in the targeted food. This treatment is used to preserve food, reduce the risk of food borne illness, prevent the spread of invasive pests, and delay or eliminate sprouting or ripening. Irradiated food does not become radioactive. The radiation can be emitted by a radioactive substance or generated electrically.

Irradiation is also used for non-food applications, such as medical devices.

Cobalt-60 irradiation facility is used to test irradiation as a tool in the battle for food safety.

The international Radura logo, used to show a food has been treated with ionizing radiation.

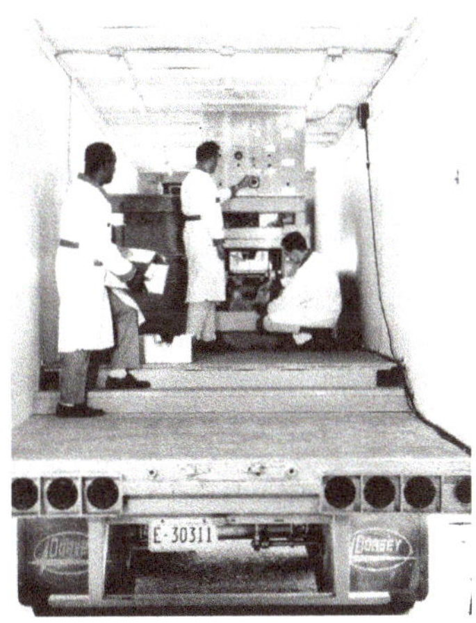

A portable, trailer-mounted food irradiation machine, circa 1968

A large amount of independent research has confirmed irradiation to be safe. One family of chemicals is uniquely formed by irradiation, and this product is nontoxic. When irradiating food, all other chemicals occur in a lower or comparable frequency to other food processing techniques. Others criticize irradiation because of confusion with radioactive contamination or because of negative impressions of the nuclear industry.

Food irradiation is permitted by over 60 countries, with about 500,000 metric tons of food annually processed worldwide. The regulations that dictate how food is to be irradiated, as well as the food allowed to be irradiated, vary greatly from country to country. In Austria, Germany, and many other countries of the European Union only dried herbs, spices, and seasonings can be processed with irradiation and only at a specific dose, while in Brazil all foods are allowed at any dose.

Uses

Irradiation is used to reduce or eliminate the risk of food born illnesses. Depending on the dose, some or all of the pathogenic organism s, microorganisms, bacteria, and viruses present are destroyed, slowed down, or rendered incapable of reproduction. Some foods are irradiated at sufficient doses to ensure that the product is sterilized and does not add any spoilage or pathogenic microorganisms into the final product.

Irradiation is used reduce post harvest losses. It reduces the pathogens in the food and slows down the speed at which enzymes produced by the food can change the food and therefore slows the spoilage, ripening, and or sprouting of the food.

Irradiation cannot revert spoiled or over ripened food to a fresh state. If this food was processed by irradiation, spoilage would cease and ripening would slow down, yet the irradiation would not destroy the toxins or repair the texture, color, or taste of the food.

Food is irradiated to prevent the spread of foreign invasive species across national boundaries. The pests are sterilized or destroyed when the food is treated by low doses of irradiation. This also allows food to pass quickly through quarantine and avoid spoilage.

Public Perception and Impact

Irradiation has been approved by the FDA for over 50 years, but the only major growth area for the commercial sale of irradiated foods for human consumption is fruits and vegetables that are irradiated to kill insects for the purpose of quarantine. In the early 2000s in the US, irradiated meat was common at some grocery stores, but because of lack of consumer demand it is no longer common. Because consumer demand for irradiated food is low, reducing the spoilage between manufacture and consumer purchase and reducing the risk of food borne illness is currently not sufficient incentive for most manufactures to supplement their process with irradiation.

It is widely believed that consumer perception of foods treated with irradiation is more negative than those processed by other means, although some industry studies indicate the number of consumers concerned about the safety of irradiated food has decreased in the last 10 years to levels comparable to those of people concerned about food additives and preservatives. "These irradiated foods are not less safe than others," Dr. Tarantino said, "and the doses are effective in reducing the level of disease-causing micro-organisms." "People think the product is radioactive," said Harlan Clemmons, president of Sadex, a food irradiation company based in Sioux City, Iowa.

Some common concerns about food irradiation include the impact of irradiation on food chemistry, as well as the cumulative impacts of the irradiation process and the indirect effects of irradiation becoming a prevalent in the food handling process. Irradiation reduces the risk of infection and spoilage, does not make food radioactive, and the food is shown to be safe, but it does cause chemical reactions that alter the food and therefore alters the chemical makeup, nutritional content, and the sensory qualities of the food. Some of the potential secondary impacts of irradiation are hypothetical, while others are demonstrated. These effects include cumulative impacts to pathogens, people, and the environment due to the reduction of food quality, the transpiration and storage of radioactive goods, and destruction of pathogens. People are also concerned about the indirect effects of irradiation, such as the way irradiation changes the way we relate to food and how irradiation changes the food production and shipping industries. Because of these concerns and the increased cost of irradiated foods, there is not a widespread public demand for the irradiation of foods for human consumption.

Immediate Effects

The radiation source supplies energetic particles or waves (photons). As these waves/particles pass through a target material they collide with other particles. Around the sites of these collisions chemical bonds are broken, creating short lived radicals (e.g. the hydroxyl radical, the hydrogen atom and solvated electrons). These radicals cause further chemical changes by bonding with and or stripping particles from nearby molecules. When collisions damage DNA or RNA, effective reproduction becomes unlikely, also when collisions occur in cells, cell division is often suppressed.

Irradiation (within the accepted energy limits, as 10 MeV for electrons, 5 MeV for X-rays [US 7.5 MeV] and gamma rays from Cobalt-60) can not make food radioactive, but it does produce radiolytic products, and free radicals in the food. A few of these products are unique, but not considered dangerous.

Irradiation can also alter the nutritional content and flavor of foods, much like cooking. The scale of these chemical changes is not unique. Cooking, smoking, salting, and other less novel techniques, cause the food to be altered so drastically that its original nature is almost unrecognizable, and must be called by a different name. Storage of food also causes dramatic chemical changes, ones that eventually lead to deterioration and spoilage.

Misconceptions

A major concern is that irradiation might cause chemical changes that are harmful to the consumer. Several national expert groups and two international expert groups evaluated the available data and concluded that any food at any dose is wholesome and safe to consume as long as it remains palatable and maintains its technical properties (e.g. feel, texture, or color).

Irradiated food does not become radioactive, just as an object exposed to light does not start producing light. Radioactivity is the ability of a substance to emit high energy particles. When particles hit the target materials they may free other highly energetic particles. This ends shortly after the end of the exposure, much like objects stop reflecting light when the source is turned off and warm objects emit heat until they cool down but do not continue to produce their own heat. To modify a material so that it keeps emitting radiation (induce radiation) the atomic cores (nucleus) of the atoms in the target material must be modified.

It is impossible for food irradiators to induce radiation into a product. Irradiators emit electrons or photons and the radiation is intrinsically radiated at precisely known strengths (wavelengths for photons, and speeds for electrons). These radiated particles at these strengths can never be strong enough to modify the nucleus of the targeted atom in the food, regardless of how many particles hit the target material, and radioactivity can not be induced without modifying the nucleus.

Chemical Changes

Compounds known as free radicals form when food is irradiated. Most of these are oxidizers (i.e., accept electrons) and some react very strongly. According to the free-radical theory of aging excessive amounts of these free radicals can lead to cell injury and cell death, which may contribute to many diseases. However, this generally relates to the free radicals generated in the body, not the free radicals consumed by the individual, as much of these are destroyed in the digestive process.

When fatty acids are irradiated, a family of compounds called 2-alkylcyclobutanones (2-ACBs) are produced. These are thought to be unique radiolytic products. Most of the substances found in irradiated food are also found in food that has been subjected to other food processing treatments, and are therefore not unique. Furthermore, the quantities in which they occur in irradiated food are lower or similar to the quantities formed in heat treatments.

The radiation doses to cause toxic changes are much higher than the doses used to during irradiation, and taking into account the presence of 2-ACBs along with what is known of free radicals, these results lead to the conclusion that there is no significant risk from radiolytic products.

Food Quality

Because of the extent of the chemical reactions, changes to the foods quality after irradiation are inevitable. The nutritional contents, as well as the sensory qualities (taste, appearance, and texture) are impacted by irradiation. Irradiation changes the sensory qualities and reduces the content of some nutrients in foods. Irradiation also improves the shelf stability of some sensory qualities and nutrients. For these components more of the original taste, texture, appearance, and nutrients get to the final consumer. The changes in quality and nutrition vary greatly from food to food.

There has been low level gamma irradiation that has been attempted on arugula, spinach, cauliflower, ash gourd, bamboo shoots, coriander, parsley, and watercress. There has been limited information, however, regarding the physical, chemical and/or bioactive properties and the shelf life on these minimally processed vegetables.

Because the nutrational content changes after irradiation food advocacy groups consider labeling irradiated food raw as misleading. However, the degradation of vitamins caused by irradiation is similar or even less than the loss caused by other food preservation processes. Other processes like chilling, freezing, drying, and heating also result in some vitamin loss.

The changes in the flavor of fatty foods like meats, nuts and oils are sometimes noticeable, while the changes in lean products like fruits and vegetables are less so. Some studies by the irradiation industry show that for some properly treated fruits and vegetables irradiation is seen by consumers to improve the sensory qualities of the product compared to untreated fruits and vegetables.

Quality Impact on Minimally Processed Vegetables

Watercress (*Nasturtium Officinale*) is a rapidly growing aquatic or semi aquatic perennial plant. It contains health promoting phytochemicals endowed in therapeutic properties. Because chemical agents do not provide efficient microbial reductions, watercress has been tested with gamma irradiation treatment in order to improve both safety and the shelf life of the product. It is traditionally used on horticultural products to prevent sprouting and post-packaging contamination, delay post-harvest ripening, maturation and senescence.

In a Food Chemistry food journal, scientists studied the suitability of gamma irradiation of 1, 2, and 5 kGy for preserving quality parameters of the fresh cut watercress at around 4 degrees Celsius for 7 days. They determined that a 2 kGy dose of irradiation was the dose that contained most similar qualities to non-stored control samples, which is one of the goals of irradiation. 2 kGy preserved high levels of reducing sugars and favoured PUFA; while samples of the 5 kGy dose revealed high contents of sucrose and MUFA. Both cases the watercress samples obtained healthier fatty acids profiles. However, a 5kGy dose better preserved the antioxidant activity and total flavonoids.

Long Term Impacts

If the majority of food was irradiated at high enough levels to significantly decrease its nutritional content, there would be an increased risk to develop illnesses that are nutritional based if additional steps, such as changes in eating habits, where not taken to mitigate this. Furthermore, for at least 3 studies on cats, the consumption of irradiated food was associated with a loss of tissue in the myelin sheath, leading to reversible paralysis. Researchers suspect that reduced levels of vitamin A and high levels of free radicals may be the cause. This effect is thought to be specific to cats and has not been reproduced in any other animal. To produce these effects the cats were fed solely on food that was irradiated at a dose at least five times higher than the maximum allowable dose.

It may seem reasonable to assume that irradiating food might lead to radiation-tolerant strains, similar to the way that strains of bacteria have developed resistance to antibiotics. Bacteria develop a resistance to antibiotics after an individual uses antibiotics repeatedly. Much like pasteurization plants products that pass through irradiation plants are processed once, and are not processed and reprocessed. Cycles of heat treatment have been shown to produce heat tolerant bacteria, yet no problems have appeared so far in pasteurization plants. Furthermore, when the irradiation dose is chosen to target a specific species of microbe, it is calibrated to doses several times the value required to target the species. This ensures that the process randomly destroys all members of a target species. Therefore, the more irradiation tolerant members of the target species are not given any evolutionary advantage. Without evolutionary advantage selection does not occur. As to the irradiation process directly producing mutations that lead to more virulent, radiation resistant, strains the European Commission's Scientific Committee on Food found that there is no evidence, on the contrary, irradiation has been found to cause loss of virulence and infectivity as mutants are usually less competitive and less adapted."

Misconceptions

The argument is made that there is a lack of long-term studies, and therefore the safety of irradiated food is not scientifically proven in spite of the fact that hundreds of animal feeding studies of irradiated food, including multigenerational studies, have been performed since 1950. Endpoints investigated have included subchronic and chronic changes in metabolism, histopathology, and function of most systems; reproductive effects; growth; teratogenicity; and mutagenicity. A large number of studies have been performed; meta-studies have supported the safety of irradiated food.

The below experiments are cited by food irradiation opponents, but could be either not verified in later experiments, could not be clearly attributed to the radiation effect, or could be attributed to an inappropriate design of the experiment etc.

- India's National Institute of Nutrition (NIN) found an elevated rate of cells with more than one set of genes (Polyploidy) in humans and animals when fed wheat that was irradiated recently (within 12 weeks). Upon analysis scientist determined that the techniques used by NIN allowed for too much human error and statistical variation, therefore the results where unreliable. After multiple studies by independent agencies and scientists no correlation between polyploidy and irradiation of food could be found.

Indirect Effects of Irradiation

The indirect effects of irradiation are the concerns and benefits of irradiation that are related to how making food irradiation a common process will change the world, with emphasis on the system of food production.

If irradiation was to become common in the food handling process there would be a reduction of the prevalence of foodborne illness and potentially the eradication of specific pathogens. However, multiple studies suggest that an increased rate of pathogen growth may occur when irradiated food is cross-contaminated with a pathogen, as the competing spoilage organisms are no longer present. This being said, cross contamination itself becomes less prevalent with an increase in usage of irradiated foods.

The ability to remove bacterial contamination through post-processing by irradiation may reduce the fear of mishandling food which could cultivate a cavalier attitude toward hygiene and result in contaminants other than bacteria. However, concerns that the pasteurization of milk would lead to increased contamination of milk were prevalent when mandatory pasteurization was introduced, but these fears never materialized after adoption of this law. Therefore, it is unlikely for irradiation to cause an increase of illness due to nonbacteria-based contamination.

Treatment

Up to the point where the food is processed by irradiation, the food is processed in the same way as all other food. To treat the food, they are exposed to a radioactive source, for a set period of time to achieve a desired dose. Radiation may be emitted by a radioactive substance, or by X-ray and electron beam accelerators. Special precautions are taken to ensure the food stuffs never come in contact with the radioactive substances and that the personnel and the environment are protected from exposure radiation. Irradiation treatments are typically classified by dose (high, medium, and low), but are sometimes classified by the effects of the treatment (radappertization, radicidation and radurization). Food irradiation is sometimes referred to as "cold pasteurization" or "electronic pasteurization" because ionizing the food does not heat the food to high temperatures during the process, and the effect is similar to heat pasteurization. The term "cold pasteurization" is controversial because the term may be used to disguise the fact the food has been irradiated and pasteurization and irradiation are fundamentally different processes.

Treatment costs vary as a function of dose and facility usage. A pallet or tote is typically exposed for several minutes to hours depending on dose. Low-dose applications such as disinfestation of fruit range between US$0.01/lbs and US$0.08/lbs while higher-dose applications can cost as much as US$0.20/lbs.

Process

Typically, when the food is being irradiated, pallets of food are exposed to a source of radiation for a specific time. Dosimeters are embedded in the pallet (at various locations) of food to determine what dose was achieved. Most irradiated food is processed by gamma irradiation., however the usage of electron beam and X-ray is becoming more popular as well . Special precautions are taken because gamma rays are continuously emitted by the radioactive material. In most designs, to nullify the effects of radiation, the radioisotope is lowered into a water-filled storage pool, which absorbs the radiation but does not become radioactive. This allows pallets of the products to be added and removed from the irradiation chamber and other maintenance to be done. Sometimes movable shields are used to reduce radiation levels in areas of the irradiation chamber instead of submerging the source. For x ray and electron irradiation these precautions are not necessary as the source of the radiation can be switched off.

For x-ray, gamma ray and electron irradiation, shielding is required when the foods are being irradiated. This is done to protect workers and the environment outside of the chamber from radiation exposure. Typically permanent or movable shields are used. In some gamma irradiators the radioactive source is under water at all times, and the hermetically sealed product is lowered into the water. The water acts as the shield in this application. Because of the lower penetration depth of electron irradiation, treatment to entire industrial pallets or totes is not possible.

Dosimetry

The radiation absorbed dose is the amount energy absorbed per unit weight of the target material. Dose is used because, when the same substance is given the same dose, similar changes are observed in the target material. The SI unit for dose is grays (Gy or J/kg). Dosimeters are used to measure dose, and are small components that, when exposed to ionizing radiation, change measurable physical attributes to a degree that can be correlated to the dose received. Measuring dose (dosimetry) involves exposing one or more dosimeters along with the target material.

For purposes of legislation doses are divided into low (up to 1 kGy), medium (1 kGy to 10 kGy), and high-dose applications (above 10 kGy). High-dose applications are above those currently permitted in the US for commercial food items by the FDA and other regulators around the world. Though these doses are approved for non commercial applications, such as sterilizing frozen meat for NASA astronauts (doses of 44 kGy) and food for hospital patients.

Applications By Overall Average Dose		
	Application	**Dose (kGy)**
Low dose (up to 1 kGy)	Inhibit sprouting[a]	0.03-0.15 kGy
	Delay fruit ripening	0.03-0.15 kGy
	Stop insect/parasite infestations[b]	0.07-1.00 kGy
Medium dose (1 kGy to 10 kGy)	Delay spoilage of meat[c]	1.50–3.00 kGy
	Reduce risk of pathogens in meat[c]	3.00–7.00 kGy
	Increase sanitation[d] of spices	10.00 kGy
High dose (above 10 kGy)	Sterilization[e] of packaged meat[c]	25.00-70.00 kGy
	Increase juice yield	
	Improve re-hydration	

Technology

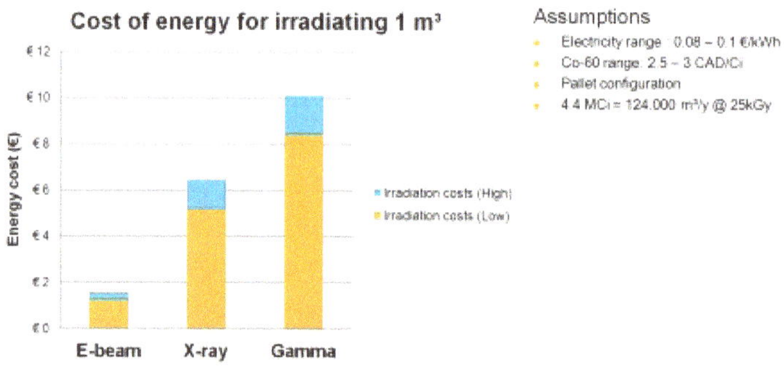

Efficiency illustration of the different radiation technologies (electron beam, X-ray, gamma rays)

Electron irradiation uses electrons accelerated in an electric field to a velocity close to the speed of light. Electrons have a charge, and therefore do not penetrate the product beyond a few centimeters, depending on product density.

Gamma irradiation involves exposing the target material to packets of light (photons) that are highly energetic (Gamma rays). A radioactive material (radioisotopes) is used as the source for the gamma rays. Gamma irradiation is the standard because the deeper penetration of the gamma rays enables administering treatment to entire industrial pallets or totes (reducing the need for material handling) and it is significantly less expensive than using a X-ray source. Generally cobalt-60 is used as a radioactive source for gamma irradiation. Cobalt-60 is bred from cobalt-59 using neutron irradiation in specifically designed nuclear reactors. In limited applications caesium-137, a less costly alternative recovered during the processing of spent nuclear fuel, is used as a radioactive source. Insufficient quantities are available for large scale commercial use. An incident where water-soluble caesium-137 leaked into the source storage pool requiring NRC intervention has led to near elimination of this radioisotope outside of military applications.

Irradiation by X-ray is similar to irradiation by gamma rays in that less energetic packets of light (X-rays) are used. X-rays are generated by colliding accelerated electrons with a dense material (this process is known as bremsstrahlung-conversion), and therefore do not necessitate the use of radioactive materials. X-rays ability to penetrate the target is similar to gamma irradiation. X-ray machines produce better dose uniformity than Gamma irradiation but require much more electricity as only as much as 12% of the input energy is converted into X-rays.

Cost

The cost of food irradiation is influenced by dose requirements, the food's tolerance of radiation, handling conditions, i.e., packaging and stacking requirements, construction costs, financing arrangements, and other variables particular to the situation. Irradiation is a capital-intensive technology requiring a substantial initial investment, ranging from $1 million to $5 million. In the case of large research or contract irradiation facilities, major capital costs include a radiation source, hardware (irradiator, totes and conveyors, control systems, and other auxiliary equipment), land (1 to 1.5 acres), radiation shield, and warehouse. Operating costs include salaries (for fixed and variable labor), utilities, maintenance, taxes/insurance, cobalt-60 replenishment, general utilities, and miscellaneous operating costs.

Standards & Regulations

The Codex Alimentarius represents the global standard for irradiation of food, in particular under the WTO-agreement. Member states are free to convert those standards into national regulations at their discretion, therefore regulations about irradiation differ from country to country.

The United Nations Food and Agricultural Organization (FAO) has passed a motion to commit member states to implement irradiation technology for their national phytosanitary programs; the General assembly of the International Atomic Energy Agency (IAEA) has urged wider use of the irradiation technology.

Labeling

The Radura symbol, as required by U.S. Food and Drug Administration regulations to show a food has been treated with ionizing radiation.

The provisions of the Codex Alimentarius are that any "first generation" product must be labeled "irradiated" as any product derived directly from an irradiated raw material; for ingredients the provision is that even the last molecule of an irradiated ingredient must be listed with the ingredients even in cases where the unirradiated ingredient does not appear on the label. The RADURA-logo is optional; several countries use a graphical version that differs from the Codex-version. The suggested rules for labeling is published at CODEX-STAN – 1 (2005), and includes the usage of the Radura symbol for all products that contain irradiated foods. The Radura symbol is not a designator of quality. The amount of pathogens remaining is based upon dose and the original content and the dose applied can vary on a product by product basis.

The European Union follows the Codex's provision to label irradiated ingredients down to the last molecule of irradiated food. The European Community does not provide for the use of the Radura logo and relies exclusively on labeling by the appropriate phrases in the respective languages of the Member States. The European Union enforces its irradiation labeling laws by requiring its member countries to perform tests on a cross section of food items in the market-place and to report to the European Commission. The results are published annually in the OJ of the European Communities.

The US defines irradiated foods as foods in which the irradiation causes a material change in the food, or a material change in the consequences that may result from the use of the food. Therefore, food that is processed as an ingredient by a restaurant or food processor is exempt from the labeling requirement in the US. This definition is not consistent with the Codex Alimentarius. All irradiated foods must bear a slightly modified Radura symbol at the point of sale and use the term "irradiated" or a derivative there of, in conjunction with explicit language describing the change in the food or its conditions of use.

Food Safety

In 2003, the Codex Alimentarius removed any upper dose limit for food irradiation as well as clearances for specific foods, declaring that all are safe to irradiate. Countries such as Pakistan and Brazil have adopted the Codex without any reservation or restriction. Other countries, including New Zealand, Australia, Thailand, India, and Mexico, have permitted the irradiation of fresh fruits for fruit fly quarantine purposes, amongst others.

Standards that describe calibration and operation for radiation dosimetry, as well as procedures to relate the measured dose to the effects achieved and to report and document such results, are maintained by the American Society for Testing and Materials (ASTM international) and are also available as ISO/ASTM standards.

All of the rules involved in processing food are applied to all foods before they are irradiated.

United States

In the United States, each new food is approved separately with a guideline specifying a maximum dosage; in case of quarantine applications the minimum dose is regulated. Packaging materials containing the food processed by irradiation must also undergo approval. Food irradiation in the United States is primarily regulated by the FDA since it is considered a food additive. The United States Department of Agriculture (USDA) amends these rules for use with meat, poultry, and fresh fruit.

The United States Department of Agriculture (USDA) has approved the use of low-level irradiation as an alternative treatment to pesticides for fruits and vegetables that are considered hosts to a number of insect pests, including fruit flies and seed weevils. Under bilateral agreements that allows less-developed countries to earn income through food exports agreements are made to allow them to irradiate fruits and vegetables at low doses to kill insects, so that the food can avoid quarantine.

The U.S. Food and Drug Administration (FDA) and the USDA have approved irradiation of the following foods and purposes:

- Packaged refrigerated or frozen red meat — to control pathogens (E. Coli O157:H7 and Salmonella), and to extend shelf life.
- Packaged poultry — control pathogens (Salmonella and Camplylobacter).
- Fresh fruits, vegetables and grains — to control insects and inhibit growth, ripening and sprouting.
- Pork — to control trichinosis.
- Herbs, spices and vegetable seasonings — to control insects and microorganisms.
- Dry or dehydrated enzyme preparations — to control insects and microorganisms.
- White potatoes — to inhibit sprout development.
- Wheat and wheat flour — to control insects.
- Loose or bagged fresh iceberg lettuce and spinach

European Union

European law dictates that no foods other than dried aromatic herbs, spices and vegetable seasonings are permitted for the application of irradiation. However, any Member State is permitted to maintain previous clearances that are in categories that the EC's Scientific Committee on Food (SCF) had previously approved, or add clearance granted to other Member States. Presently, Belgium, Czech Republic, France, Italy, Netherlands, Poland, and the United Kingdom have adopted such provisions. Before individual items in an approved class can be added to the approved list, studies into the toxicology of each of such food and for each of the proposed dose ranges are requested. It also states that irradiation shall not be used "as a substitute for hygiene or health practices or good manufacturing or agricultural practice". These regulations only govern food irradiation in consumer products to allow irradiation to be used for patients requiring sterile diets.

Because of the Single Market of the EC any food, even if irradiated, must be allowed to be marketed in any other Member State even if a general ban of food irradiation prevails, under the condition that the food has been irradiated legally in the state of origin. Furthermore, imports into the EC are possible from third countries if the irradiation facility had been inspected and approved by the EC and the treatment is legal within the EC or some Member state.

Australia

Australia banned irradiated cat food after a national scare where cats suffered from paralyzation after eating a specific brand of highly irradiated catfood for an extended period of time. The suspected culprit was malnutrition from consuming food depleted of Vitamin A by the irradiation process. The incident was linked only to a single batch of one brand's product and no illness was linked to any of that brand's other irradiated batches of the same product or to any other brand of irradiated cat food. This, along with incomplete evidence indicating that the cat food was not sufficiently depleted of Vitamin A makes irradiation a less likely cause. Further research has been able to experimentally induce the paralyzation of cats by via Vitamin A deficiency by feeding highly irradiated food.

Nuclear Safety & Security

Interlocks and safeguards are mandated to minimize this risk. There have been radiation related accidents, deaths, and injury at such facilities, many of them caused by operators overriding the safety related interlocks. In a radiation processing facility, radiation specific concerns are supervised by special authorities, while "Ordinary" occupational safety regulations are handled much like other businesses.

The safety of irradiation facilities is regulated by the United Nations International Atomic Energy Agency and monitored by the different national Nuclear Regulatory Commissions. The regulators enforce a safety culture that mandates that all incidents that occur are documented and thoroughly analyzed to determine the cause and improvement potential. Such incidents are studied by personnel at multiple facilities, and improvements are mandated to retrofit existing facilities and future design.

In the US the Nuclear Regulatory Commission (NRC) regulates the safety of the processing facility, and the United States Department of Transportation (DOT) regulates the safe transport of the radioactive sources.

Irradiated Food Supply

Authorities in some countries use tests that can detect the irradiation of food items to enforce labeling standards and to bolster consumer confidence. The European Union monitors the market to determine the quantity of irradiated foods, if irritated foods are labeled as irradiated, and if the irritation is performed at approved facilities.

Irradiation of fruits and vegetables to prevent the spread of pest and diseases across borders has been increasing globally. In 2010, 18446 tonnes of fruits and vegetables were irradiated in six countries for export quarantine control; the countries follow: Mexico (56.2%), United States (31.2%), Thailand (5.18%), Vietnam (4.63%), Australia (2.69%), and India (0.05%). The three types of fruits irradiated the most were guava (49.7%), sweet potato(29.3%) and sweet lime (3.27%).

In total, 103 000 tonnes of food products were irradiated on mainland United States in 2010. The three types of foods irradiated the most were spices (77.7%), fruits and vegetables (14.6%) and meat and poultry (7.77%). 17 953 tonnes of irradiated fruits and vegetables were exported to the mainland United States. Mexico, the United States' state of Hawaii, Thailand, Vietnam and India export irradiated produce to the mainland U.S. Mexico, followed by the United States' state of Hawaii, is the largest exporter of irradiated produce to the mainland U.S.

In total, 6 876 tonnes of food products were irradiated in European Union countries in 2013; mainly in four member state countries: Belgium (49.4%), the Netherlands (24.4%), Spain (12.7%)

and France (10.0%). The two types of foods irradiated the most were frog legs (46%), and dried herbs and spices (25%). There has been a decrease of 14% in the total quantity of products irradiated in the EU compared to the previous year 2012 (7 972 tonnes)

Timeline of the History of Food Irradiation

- 1895 Wilhelm Conrad Röntgen discovers X-rays ("bremsstrahlung", from German for radiation produced by deceleration)
- 1896 Antoine Henri Becquerel discovers natural radioactivity; Minck proposes the therapeutic use
- 1904 Samuel Prescott describes the bactericide effects Massachusetts Institute of Technology (MIT)
- 1906 Appleby & Banks: UK patent to use radioactive isotopes to irradiate particulate food in a flowing bed
- 1918 Gillett: U.S. Patent to use X-rays for the preservation of food
- 1921 Schwartz describes the elimination of Trichinella from food
- 1930 Wuest: French patent on food irradiation
- 1943 MIT becomes active in the field of food preservation for the U.S. Army
- 1951 U.S. Atomic Energy Commission begins to co-ordinate national research activities
- 1958 World first commercial food irradiation (spices) at Stuttgart, Germany
- 1970 Establishment of the International Food Irradiation Project (IFIP), headquarters at the Federal Research Centre for Food Preservation, Karlsruhe, Germany
- 1980 FAO/IAEA/WHO Joint Expert Committee on Food Irradiation recommends the clearance generally up to 10 kGy "overall average dose"
- 1981/1983 End of IFIP after reaching its goals
- 1983 Codex Alimentarius General Standard for Irradiated Foods: any food at a maximum "overall average dose" of 10 kGy
- 1984 International Consultative Group on Food Irradiation (ICGFI) becomes the successor of IFIP
- 1998 The European Union's Scientific Committee on Food (SCF) voted "positive" on eight categories of irradiation applications
- 1997 FAO/IAEA/WHO Joint Study Group on High-Dose Irradiation recommends to lift any upper dose limit
- 1999 The European Union issues Directives 1999/2/EC (framework Directive) and 1999/3/EC (implementing Directive) limiting irradiation a positive list whose sole content is one of

the eight categories approved by the SFC, but allowing the individual states to give clearances for any food previously approved by the SFC.

- 2000 Germany leads a veto on a measure to provide a final draft for the positive list.
- 2003 Codex Alimentarius General Standard for Irradiated Foods: no longer any upper dose limit
- 2003 The SCF adopts a "revised opinion" that recommends against the cancellation of the upper dose limit.
- 2004 ICGFI ends
- 2011 The successor to the SFC, European Food Safety Authority (EFSA), reexamines the SFC's list and makes further recommendations for inclusion.

Pascalization

Pascalization, bridgmanization, or high pressure processing (HPP), is a method of preserving and sterilizing food, in which a product is processed under very high pressure, leading to the inactivation of certain microorganisms and enzymes in the food. The technique was named after Blaise Pascal, a French scientist of the 17th century whose work included detailing the effects of pressure on fluids. During pascalization, more than 50,000 pounds per square inch (340 MPa, 3.4 kbar) may be applied for around fifteen minutes, leading to the inactivation of yeast, mold, and bacteria. Pascalization is also known as bridgmanization, named for physicist Percy Williams Bridgman.

Uses

Pascalization stops chemical activity caused by microorganisms that play a role in the deterioration of foods. The treatment occurs at low temperatures and does not include the use of food additives. From 1990, some juices, jellies, and jams have been preserved using pascalization in Japan. The technique is now used there to preserve fish and meats, salad dressing, rice cakes, and yogurts. HPP is now being used to preserve fruit and vegetable smoothies and other products such as meat for sale in the UK,. An early use of pascalization in the United States was to treat guacamole. It did not change the guacamole's taste, texture, or color, but the shelf life of the product increased to thirty days, from three days without the treatment. However, some treated foods still require cold storage because pascalization does not stop all enzyme activity caused by proteins, some of which affects shelf life.

History

Late 1800s

Experiments into the effects of pressure on microorganisms have been recorded as early as 1884, and successful experiments since 1897. In 1899, B. H. Hite was the first to conclusively demonstrate the inactivation of microorganisms by pressure. After he reported the effects of high pressure on microorganisms, and reports on the effects of pressure on foods quickly followed. Hite tried to prevent milk from spoiling, and his work showed that microorganisms can be deactivated by subjecting it to high pressure. He also mentioned some advantages of pressure-treating foods, such as the lack of antiseptics and no change in taste.

Hite said that, since 1897, a chemist at the West Virginia Agricultural Experimental Station had been studying the relationship between pressure and the preservation of meats, juices, and milk. Early experiments involved inserting a large screw into a cylinder and keeping it there for several days, but this did not have any effect in stopping the milk from spoiling. Later, a more powerful apparatus was able to subject the milk to higher pressures, and the treated milk was reported to stay sweeter for 24–60 hours longer than untreated milk. When 90 short tons (82 t) of pressure was applied to samples of milk for one hour, they stayed sweet for one week. Unfortunately, the device used to induce pressure was later damaged when researchers tried to test its effects on other products.

Experiments were also performed with anthrax, typhoid, and tuberculosis, which was a potential health risk for the researchers. Indeed, before the process was improved, one employee of the Experimental Station became ill with typhoid fever.

The process that Hite reported on was not feasible for widespread use and did not always completely sterilize the milk. While more extensive investigations followed, the original study into milk was largely discontinued due to concerns over its effectiveness. Hite mentioned "certain slow changes in the milk" related to "enzymes that the pressure could not destroy."

Early 1900s

Hite et al. released a more detailed report on pressure sterilization in 1914, which included the number of microorganisms that remained in a product after treatment. Experiments were conducted on various other foods, including fruits, fruit juices, and some vegetables. They were met with mixed success, similar to the results obtained from the earlier tests on milk. While some foods were preserved, others were not, possibly due to bacterial spores that had not been killed.

Hite's 1914 investigation led to other studies into the effect of pressure on microorganisms. In 1918, a study published by W. P. Larson et al. was intended to help advance vaccines. This report showed that bacterial spores were not always inactivated by pressure, while vegetative bacteria were usually killed. Larson et al.'s investigation also focused on the use of carbon dioxide, hydrogen, and nitrogen gas pressures. Carbon dioxide was found to be the most effective of the three at inactivating microorganisms.

Late 1900s–Today

Around 1970, researchers renewed their efforts in studying bacterial spores after it was discovered that using moderate pressures was more effective than using higher pressures. These spores, which caused a lack of preservation in the earlier experiments, were inactivated faster by moderate pressure, but in a manner different from what occurred with vegetative microbes. When subjected to moderate pressures, bacterial spores germinate, and the resulting spores are easily killed using pressure, heat, or ionizing radiation. If the amount of initial pressure is increased, conditions are not ideal for germination, so the original spores must be killed instead. However, using moderate pressure does not always work, as some bacterial spores are more resistant to germination under pressure and a small portion of them will survive. A preservation method using both pressure and another treatment (such as heat) to kill spores has not yet been reliably achieved. Such a technique would allow for wider use of pressure on food and other potential advancements in food preservation.

Research into the effects of high pressures on microorganisms was largely focused on deep-sea organisms until the 1980s, when advancements in ceramic processing were made. This resulted in the production of machinery that allowed for processing foods at high pressures at a large scale, and generated some interest in the technique, especially in Japan. Although commercial products preserved by pascalization first emerged in 1990, the technology behind pascalization is still being perfected for widespread use. There is now higher demand for minimally processed products than in previous years, and products preserved by pascalization have seen commercial success despite being priced significantly higher than products treated with standard methods.

In the early 21st century, it was discovered that pascalization can separate the meat of shellfish from their shells. Lobsters, shrimp, crabs, etc. may be pascalized, and afterwards their raw meat will simply and easily slide whole right out of the cracked shell.

Process

In pascalization, food products are sealed and placed into a steel compartment containing a liquid, often water, and pumps are used to create pressure. The pumps may apply pressure constantly or intermittently. The application of high hydrostatic pressures (HHP) on a food product will kill many microorganisms, but the spores of some bacteria may need to be separately treated with acid to prevent their reproduction. Pascalization works especially well on acidic foods, such as yogurts and fruits, because pressure-tolerant spores are not able to live in environments with low pH levels. The treatment works equally well for both solid and liquid products.

During pascalization, the food's proteins are denatured, hydrogen bonds are fortified, and noncovalent bonds in the food are disrupted, while the product's main structure remains intact. Because pascalization is not heat-based, covalent bonds are not affected, causing no change in the food's taste. High hydrostatic pressure can affect muscle tissues by increasing the rate of lipid oxidation, which in turn leads to poor flavor and decreased health benefits.

Because hydrostatic pressure is able to act quickly and evenly on food, neither the size of a product's container nor its thickness play a role in the effectiveness of pascalization. There are several side effects of the process, including a slight increase in a product's sweetness, but pascalization does not greatly affect the nutritional value, taste, texture, and appearance. As a result, high pressure treatment of foods is regarded as a "natural" preservation method, as it does not use chemical preservatives.

Criticism

Anurag Sharma, a geochemist, James Scott, a microbiologist, and others at the Carnegie Institution of Washington directly observed microbial activity at pressures in excess of 1 gigapascal. The experiments were performed up to 1.6 GPa of pressure, which is more than 16,000 times normal air pressure, or about 14 times the pressure in the deepest ocean trench.

The experiment began by depositing an *Escherichia coli* and *Shewanella oneidensis* film in a Diamond Anvil Cell (DAC). The pressure was then raised to 1.6 GPa. When raised to this pressure and kept there for 30 hours, at least 1% of the bacteria survived. The experimenters then monitored formate metabolism using in-situ Raman spectroscopy and showed that formate metabolism continued in the bacterial sample.

Moreover, 1.6 GPa is such great pressure that during the experiment the DAC turned the solution into ice-IV, a room-temperature ice. When the bacteria broke down the formate in the ice, liquid pockets would form because of the chemical reaction.

There was some skepticism of this experiment. According to Art Yayanos, an oceanographer at the Scripps Institute of Oceanography, an organism should only be considered living if it can reproduce. Another issue with the DAC experiment is that when high pressures occur, there are usually high temperatures present as well, but in this experiment there were not. This experiment was performed at room-temperature. However, the intentional lack of high temperature in the experiments isolated the actual effects of pressure on life and results clearly indicated life to be largely pressure insensitive.

Newer results from independent research groups have confirmed Sharma et al. (2002). This is a significant step that reiterates the need for a new approach to the old problem of studying environmental extremes through experiments. There is practically no debate whether microbial life can survive pressures up to 600 MPa, which has been shown over the last decade or so to be valid through a number of scattered publications.

Consumer Acceptance of Pascalization

In the consumer studies of Hightech Europe consumers mentioned more positive than negative associations descriptions for this technology showing that these products are well accepted.

Hurdle Technology

Hurdle technology is a method of ensuring that pathogens in food products can be eliminated or controlled. This means the food products will be safe for consumption, and their shelf life will be extended. Hurdle technology usually works by combining more than one approach. These approaches can be thought of as "hurdles" the pathogen has to overcome if it is to remain active in the food. The right combination of hurdles can ensure all pathogens are eliminated or rendered harmless in the final product.

Hurdle technology has been defined by Leistner (2000) as an intelligent combination of hurdles which secures the microbial safety and stability as well as the organoleptic and nutritional quality and the economic viability of food products. The organoleptic quality of the food refers to its sensory properties, that is its look, taste, smell and texture.

Examples of hurdles in a food system are high temperature during processing, low temperature during storage, increasing the acidity, lowering the water activity or redox potential, or the presence of preservatives. According to the type of pathogens and how risky they are, the intensity of the hurdles can be adjusted individually to meet consumer preferences in an economical way, without compromising the safety of the product.

Hurdles

Each hurdle aims to eliminate, inactivate or at least inhibit unwanted microorganisms. Common salt or organic acids can be used as hurdles to control microbials in food. Many natural antimicrobials such as nisin, natamycin and other bacteriocins, and essential oils derived from rosemary or thyme, also work well.

Principal hurdles used for food preservation (after Leistner, 1995)		
Parameter	**Symbol**	**Application**
High temperature	F	Heating
Low temperature	T	Chilling, freezing
Reduced water activity	a_w	Drying, curing, conserving
Increased acidity	pH	Acid addition or formation
Reduced redox potential	E_h	Removal of oxygen or addition of ascorbate
Biopreservatives		Competitive flora such as microbial fermentation
Other preservatives		Sorbates, sulfites, nitrites

"Traditionally, fermented seafood products common in Japan, provide a typical example of hurdle technology. Fermentation of sushi employs hurdles that favour growth of desirable bacteria but inhibit the growth of pathogens. The important hurdles in the early stages of fermentation are salt and vinegar. Raw fish is cured in salt (20–30%, w/w) for one month before being desalted and pickled in vinegar. The main target of these hurdles is *C. botulinum*. Growth of lactic acid bacteria during fermentation results in acid production from metabolism of added sugars and rice. The result is a pH hurdle important in controlling growth of *C. botulinum*."

Types of hurdles used for food preservation (from Ohlsson and Bengtsson, 2002)	
Type of hurdle	**Examples**
Physical	Aseptic packaging, electromagnetic energy (microwave, radio frequency, pulsed magnetic fields, high electric fields), high temperatures (blanching, pasteurization, sterilization, evaporation, extrusion, baking, frying), ionizing radiation, low temperature (chilling, freezing), modified atmospheres, packaging films (including active packaging, edible coatings), photodynamic inactivation, ultra-high pressures, ultrasonication, ultraviolet radiation
Physicochemical	Carbon dioxide, ethanol, lactic acid, lactoperoxidase, low pH, low redox potential, low water activity, Maillard reaction products, organic acids, oxygen, ozone, phenols, phosphates, salt, smoking, sodium nitrite/nitrate, sodium or potassium sulphite, spices and herbs, surface treatment agents
Microbial	Antibiotics, bacteriocins, competitive flora, protective cultures

Synergistic Effects

There can be significant synergistic effects between hurdles. For example, Gram-positive bacteria include some of the more important spoilage bacteria, such as *Clostridium*, *Bacillus* and *Listeria*. A synergistic enhancement occurs if nisin is used against these bacteria in combination with antioxidants, organic acids or other antimicrobials. Combining antimicrobial hurdles in an intelligent way means other hurdles can be reduced, yet the resulting food can have superior sensory qualities.

Biopreservation

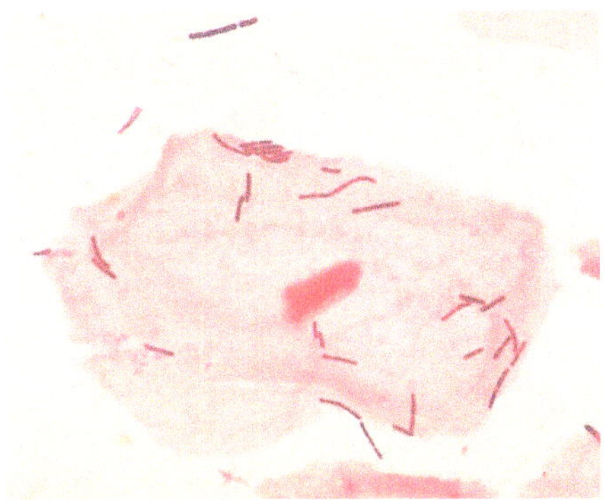

The small rods shown here are lactic acid bacteria which convert lactose and other sugars to lactic acid. The produts of their metabolism can have benign preservative effects.

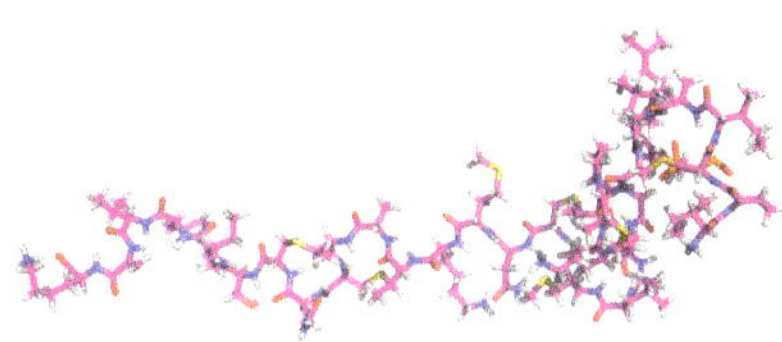

3D stick model of nisin, a particularly effective preservative produced by some lactic acid bacteria.

Biopreservation

Biopreservation is the use of natural or controlled microbiota or antimicrobials as a way of preserving food and extending its shelf life. The biopreservation of food, especially utilizing lactic acid bacteria (LAB) that are inhibitory to food spoilage microbes, has been practiced since early ages, at first unconsciously but eventually with an increasingly robust scientific foundation. Beneficial bacteria or the fermentation products produced by these bacteria are used in biopreservation to control spoilage and render pathogens inactive in food. There are a various modes of action through which microorganisms can interfere with the growth of others such as organic acid production, resulting in a reduction of pH and the antimicrobial activity of the un-dissociated acid molecules, a wide variety of small inhibitory molecules including hydrogen peroxide, etc. It is a benign ecological approach which is gaining increasing attention.

Biopreservative Agents and Modes of Action

Lactic Acid Bacteria

Of special interest are lactic acid bacteria (LAB). Lactic acid bacteria have antagonistic properties which make them particularly useful as biopreservatives. When LABs compete for nutrients, their metabolites often include active antimicrobials such as lactic and acetic acid, hydrogen peroxide, and peptide bacteriocins. Some LABs produce the antimicrobial nisin which is a particularly effective preservative. These days LAB bacteriocins are used as an integral part of hurdle technology. Using them in combination with other preservative techniques can effectively control spoilage bacteria and other pathogens, and can inhibit the activities of a wide spectrum of organisms, including inherently resistant Gram-negative bacteria." Lactic acid bacteria and propionibacteria have been extensively studies for their efficacy against spoilage causing yeasts and molds in food spoilage.

Yeast

In addition to lactic acid bacteria, yeasts also have been reported to have a biopreservation effect due to their antagonistic activities relying on the competition for nutrients, production and tolerance of high concentrations of ethanol, as well as the synthesis of a large class of antimicrobial compounds exhibiting large spectrum of activity against food spoilage microorganisms, but also against plant, animal and human pathogen.

A bacterium/yeast that is a suitable candidate for use as a biopreservative does not necessarily have to ferment the food. However, if conditions are suitable for microbial growth, then a biopreservative bacterium will compete well for nutrients with the spoilage and pathogenic bacteria in the food. As a product of its metabolism, it should also produce acids and other antimicrobial agents, particularly bacteriocins. Biopreservative bacteria, such as lactic acid bacteria, must be harmless to humans.

Bacteriophages

Bacteriophages (Greek for 'bacteria eater'), or simply phages, are viruses which infect bacteria. The majority of all bacteriophages known exhibit a double-stranded DNA genome inside the virion capsid and belong to the order of tailed phages, Caudovirales. The tailed phages can be further separated into three families: Podoviridae, which are characterized by very short tails; Myoviridae, which exhibit longer, straight and contractile tails; and Siphoviridae, which can be identified due to their long and flexible tails. Another well studied group of phages with many applications, although minor in terms of species diversity, is represented by filamentous phages which exhibit a single stranded DNA genome decorated by a helical protein layer surrounding the DNA molecule. Bacteriophages are ubiquitously distributed in nature and can also be isolated from human or animal associated microflora. They outnumber their bacterial host species by a factor of ten representing the most abundant self-replicating entities on earth with an estimated 1031 phages in total. The idea of using phages against unwanted bacteria developed shortly after their discovery. With the improvements in organic chemistry during the 1950s, exploration and development of broad spectrum antibiotics displaced interest in bacteriophage research. Several laboratories have been testing suitability of bacteriophage isolates to control certain bacterial pathogens. Significant advancements in this research have been made at the Bacteriophage Institute in Tbilisi, Georgia, where phage therapy is routinely applied in medicine research field. Today treatment of antibiotic resistant bacteria is a challenging task. Recently, research on bacteriophages has gained additional momentum in light of the identification of antibiotic-resistant pathogens of infectious diseases, wherein the application of antibiotics is not effectively working, therefore research on the application of bacteriophages is being reviewed intensely. Bacteriophages have recently received a generally recognized as safe status based on their lack of toxicity and other detrimental effects to human health for application in meat products in USA.

Phage preparations specific for *L. monocytogenes, E. coli* O157:H7, and *S. enterica* serotypes have been commercialized and approved for application in foods or as part of surface decontamination protocols.

Meat Biopreservation

In meat processing, biopreservation has been extensively studied in fermented meat products and ready to eat meat products. The use of native or artificially-introduced microbial population to improve animal health and productivity, and/or to reduce pathogenic organisms, has been termed a probiotic or competitive enhancement approach. Competitive enhancement strategies that have been developed include competitive exclusion, addition of a microbial supplement (probiotic) that improves gastrointestinal health, and adding a limiting, non-host digestible nutrient (prebiotic) that provides an existing (or introduced) commensal microbial population a competitive advantage in the gastrointestinal tract. Each of these approaches utilizes the activities of the native microbial ecosystem against pathogens by capitalizing on the natural microbial competition. Generally speaking, competitive enhancement strategies offer a natural 'green' method to reduce pathogens in the gut of food animals.

Seafood Biopreservation

Fishery products are a source of wide variety of valuable nutrients such as proteins, vitamins, minerals, omega-3 fatty acids, taurine, etc. Fishery products, however, are also associated with human intoxication and infection. Approximately 10 to 20% of food-borne illnesses are attributed to fish consumption. Changing consumer demand has driven the appeal of traditional processes applied to seafood (e.g. salting, smoking and canning) lower compared to mild technologies involving lower salt content, lower cooking temperature and vacuum packing (VP)/modified atmosphere packing (MAP). These products, designed as lightly preserved fish products (LPFP), are usually produced from fresh seafood and further processing increases risk of cross contamination. These milder treatments are usually not sufficient to destroy microorganisms, and in some cases psychrotolerant pathogenic and spoilage bacteria can develop during the extended shelf-life of LPFP. Many of these products are also eaten raw, so minimizing the presence and preventing growth of microorganisms is essential for the food quality and safety. The microbial safety and stability of food are based on an application of preservative factors called hurdles. The delicate texture and flavor of seafood are very sensitive to the decontamination technologies such as cooking, and more recent mild technologies such as pulsed light, high pressure, ozone, and ultrasound. Chemical preservatives, which are not processes but ingredients, are out of favor with consumers due to natural preservatives demand. An alternative solution that is gaining more and more attention is biopreservation technology In fish processing, biopreservation is achieved by adding antimicrobials or by increasing the acidity of the fish muscle. Most bacteria stop multiplying when the pH is less than 4.5. Traditionally, acidity has been increased by fermentation, marination or by directly adding acetic, citric or lactic acid to food products. Other preservatives include nitrites, sulphites, sorbates, benzoates and essential oils. The main reason for less documented studies for application of protective microorganisms, bacteriophages or bacteriocins on seafood products for biopreservation compared to dairy or meat products is probably that the early stages of biopreservation have occurred mainly in fermented foodstuffs that are not so developed among seafood products. The selection of potential protective bacteria in seafood products is challenging due to the fact that they need adaptation to the seafood matrix (poor in sugar and their metabolic activities should not change the initial characteristics of the product, i.e. by acidification, and not induce spoilage that could lead to a sensory rejection. Among the microbiota identified in fresh or processed seafood, LAB remains the category that offers the highest potential for direct application as a bioprotective culture or for bacteriocin production.

Commercial Applications and Products

There has been successful implementation of various phage preparation around the globe. Various applications/delivery methods in food have been developed. Bacteriophages and their endolysins can be incorporated into food systems in several ways such as spraying, dipping or immobilization, singly or in combination with other hurdles. products The phage preparation LMP-102 has been subsequently commercialized as "ListShield" Intralyx, Inc. It has been shown to be effective against 170 different strains of "*L. monocytogenes*", reducing significantly (10 to 1000-fold) the Listeria contamination when sprayed onto ready-to-eat foods, without changing the food general composition, taste, odor or color. The Intralytix company has also commercialized phage-based antimicrobial preparations like SalmoFresh and SalmoLyse for controlling *S. enterica* SalmoFresh is prepared with a cocktail of naturally occurring lytic bacteriophages that selectively and specifically kill Salmonella, including strains belonging to the most common/highly pathogenic serotypes Typhimurium, Enteritidis, Heidelberg, Newport, Hadar, Kentucky and Thompson. According to the manufacturer, SalmoFresh is specifically designed for treating foods that are at high risk for "*Salmonella*" contamination. In particular, red meat and poultry can be treated prior to grinding for significant reductions in Salmonella contamination. SalmoLyse is a reformulated phage cocktail derived from SalmoFresh in which two of the six phages in the original cocktail have been replaced. Additional bacteriophage preparations have been formulated and referenced to be used to reduce the microbial load of animals prior to slaughter and are commercially available from Omnilytics such as the BacWash product line against Salmonella Omnilytics. Another commercial application has been developed, Listex_ P100 by Micreos in The Netherlands and was granted generally recognized as safe (GRAS) status by the FDA and USDA for use in all food products.

Another significant commercial bacteriophage application is ELICOSALI, a wide range of anti-*Salmonella* and "*E. coli*" phage cocktail, for treatment of agricultural products developed by Eliava Institute at Tbilisi, Republic of Georgia Eliava Institute.

Safety

Biopreservation judiciously exploits the antimicrobial potential of naturally occurring microorganisms in food and/or their metabolites with a long history of safe use. Bacteriocins, bacteriophages and bacteriophage-encoded enzymes fall in this theory. The long and traditional role of Lactic acid bacteria on food and feed fermentations is the main factor related to the use of bacteriocins in biopreservation. LAB and their bacteriocins have been consumed unintentionally for ages, laying down a long history of safe use. Their antimicrobial spectrum of inhibition, bactericidal mode of action, relative tolerance to processing conditions (pH, NaCl, heat treatments) and the lack of toxicity towards eukaryotic cells enforces their role as biopreservatives in food. The evaluation of any new antimicrobial actives is done in meat by USDA which relies on the GRAS assessment by FDA among other suitability data.

Food Additive

Food additives are substances added to food to preserve flavor or enhance its taste and appearance. Some additives have been used for centuries; for example, preserving food by pickling (with vinegar), salting, as with bacon, preserving sweets or using sulfur dioxide as with wines. With the advent of processed foods in the second half of the twentieth century, many more additives have been introduced, of both natural and artificial origin.

Numbering

To regulate these additives, and inform consumers, each additive is assigned a unique number, termed as "E numbers", which is used in Europe for all approved additives. This numbering scheme has now been adopted and extended by the *Codex Alimentarius* Commission to internationally identify all additives, regardless of whether they are approved for use.

E numbers are all prefixed by "E", but countries outside Europe use only the number, whether the additive is approved in Europe or not. For example, acetic acid is written as E260 on products sold in Europe, but is simply known as additive 260 in some countries. Additive 103, alkanet, is not approved for use in Europe so does not have an E number, although it is approved for use in Australia and New Zealand. Since 1987, Australia has had an approved system of labelling for additives in packaged foods. Each food additive has to be named or numbered. The numbers are the same as in Europe, but without the prefix "E".

The United States Food and Drug Administration (FDA) lists these items as "generally recognized as safe" (GRAS); they are listed under both their Chemical Abstracts Service number and FDA regulation under the United States Code of Federal Regulations.

Categories

Food additives can be divided into several groups, although there is some overlap because some additives exert more than one effect. For example, salt is both a preservative as well as a flavor.

Acidulents

confer sour or acid taste. Common acidulents include vinegar, citric acid, tartaric acid, malic acid, fumaric acid, and lactic acid.

Acidity Regulators

Acidity regulators are used to control the pH of foods for stability or to affect activity of enzymes.

Anticaking Agents

Anticaking agents keep powders such as milk powder from caking or sticking.

Antifoaming and Foaming Agents

Antifoaming agents reduce or prevent foaming in foods. Foaming agents do the reverse.

Antioxidants

Antioxidants such as vitamin C are preservatives by inhibiting the degradation of food by oxygen.

Bulking Agents

Bulking agents such as starch are additives that increase the bulk of a food without affecting its taste.

Food Coloring

Colorings are added to food to replace colors lost during preparation or to make food look more attractive.

Color Retention Agents

In contrast to colorings, color retention agents are used to preserve a food's existing color.

Emulsifiers

Emulsifiers allow water and oils to remain mixed together in an emulsion, as in mayonnaise, ice cream, and homogenized milk.

Flavors

Flavors are additives that give food a particular taste or smell, and may be derived from natural ingredients or created artificially.

Flavor Enhancers

Flavor enhancers enhance a food's existing flavors. A popular example is monosodium glutamate. Some flavor enhancers have little flavor independent of the food.

Flour Treatment Agents

Flour treatment agents are added to flour to improve its color or its use in baking.

Glazing Agents

Glazing agents provide a shiny appearance or protective coating to foods.

Humectants

Humectants prevent foods from drying out.

Tracer Gas

Tracer gas allow for package integrity testing to prevent foods from being exposed to atmosphere, thus guaranteeing shelf life.

Preservatives

Preservatives prevent or inhibit spoilage of food due to fungi, bacteria and other microorganisms.

Stabilizers

Stabilizers, thickeners and gelling agents, like agar or pectin (used in jam for example) give foods a firmer texture. While they are not true emulsifiers, they help to stabilize emulsions.

Sweeteners

Sweeteners are added to foods for flavoring. Sweeteners other than sugar are added to keep the food energy (calories) low, or because they have beneficial effects for diabetes mellitus and tooth decay and diarrhea.

Thickeners

Thickening agents are substances which, when added to the mixture, increase its viscosity without substantially modifying its other properties.

Safety and Regulation

With the increasing use of processed foods since the 19th century, food additives are more widely used. Many countries regulate their use. For example, boric acid was widely used as a food preservative from the 1870s to the 1920s, but was banned after World War I due to its toxicity, as demonstrated in animal and human studies. During World War II, the urgent need for cheap, available food preservatives led to it being used again, but it was finally banned in the 1950s. Such cases led to a general mistrust of food additives, and an application of the precautionary principle led to the conclusion that only additives that are known to be safe should be used in foods. In the United States, this led to the adoption of the Delaney clause, an amendment to the Federal Food, Drug, and Cosmetic Act of 1938, stating that no carcinogenic substances may be used as food additives. However, after the banning of cyclamates in the United States and Britain in 1969, saccharin, the only remaining legal artificial sweetener at the time, was found to cause cancer in rats. Widespread public outcry in the United States, partly communicated to Congress by postage-paid postcards supplied in the packaging of sweetened soft drinks, led to the retention of saccharin, despite its violation of the Delaney clause. However, in 2000, saccharin was later found to only be carcinogenic to rats due to their unique urine chemistry.

Hyperactivity

Periodically, concerns have been expressed about a linkage between additives and hyperactivty, however “no clear evidence of ADHD was provided”.

In 2007, Food Standards Australia New Zealand published an official shoppers' guidance with which the concerns of food additives and their labeling are mediated. In the EU it can take 10 years or more to obtain approval for a new food additive. This includes five years of safety testing, followed by two years for evaluation by the European Food Safety Authority and another three years before the additive receives an EU-wide approval for use in every country in the European Union. Apart from testing and analyzing food products during the whole production process to ensure safety and compliance with regulatory standards, Trading Standards officers (in the UK) protect the public from any illegal use or potentially dangerous mis-use of food additives by performing random testing of food products.

There has been significant controversy associated with the risks and benefits of food additives. Natural additives may be similarly harmful or be the cause of allergic reactions in certain individuals. For example, safrole was used to flavor root beer until it was shown to be carcinogenic. Due to the application of the Delaney clause, it may not be added to foods, even though it occurs naturally in sassafras and sweet basil.

Micronutrients

A subset of food additives, micronutrients added in food fortification processes preserve nutrient value by providing vitamins and minerals to foods such as flour, cereal, margarine and milk which normally would not retain such high levels. Preservatives also reduce spoilage from sources such as air, bacteria, fungi, and yeast.

Standardization of its Derived Products

ISO has published a series of standards regarding the topic and these standards are covered by ICS 67.220.

Science

Many food additives absorb radiation in the ultraviolet and / or visible region of the spectrum. This absorbance can be used to determine the concentration of an additive in a sample using external calibration. However, additives may occur together and the absorbance by one could interfere with the absorbance of another. A prior separation stage is necessary and the additives are first separated by high-pressure liquid chromatography (HPLC) and then determined on-line using a UV and/or visible light detector.

Acidity Regulator

Acidity regulators, or pH control agents, are food additives added to change or maintain pH (acidity or basicity). They can be organic or mineral acids, bases, neutralizing agents, or buffering agents. Typical agents include these acids and their sodium salts: sorbic acid, acetic acid, benzoic acid, and propionic acid. Acidity regulators are indicated by their E number, such as E260 (acetic acid), or simply listed as "food acid".

Acidity regulators differ from acidulants, which are often acidic but are added to confer sour flavors. They are not intended to stabilize the food, although that can be a collateral benefit.

References

- Hwang, Andy; Huang, Lihan (31 January 2009). Ready-to-Eat Foods: Microbial Concerns and Control Measures. CRC Press. p. 88. ISBN 978-1-4200-6862-7. Retrieved 19 April 2011.
- Perdue, R (2009), "Vacuum Packaging", in Yam, K L, Encyclopedia of Packaging Technology, Wiley (published 2010), ISBN 978-0-470-08704-6
- Whaites, Eric; Roderick Cawson (2002). Essentials of Dental Radiography and Radiology. Elsevier Health Sciences. pp. 15–20. ISBN 0-443-07027-X.
- Leistner L (1995) "Principles and applications of hurdle technology" In Gould GW (Ed.) New Methods of Food Preservation, Springer, pp. 1–21. ISBN 978-0-8342-1341-8.
- Ohlsson T and Bengtsson N (2002) "The hurdle concept" Minimal Processing Technologies in the Food Industry, Woodhead Publishing, pp. 175–195. ISBN 978-1-85573-547-7.
- Ananou S, Maqueda M, Martínez-Bueno M and Valdivia E (2007) "Biopreservation, an ecological approach to improve the safety and shelf-life of foods" In: A. Méndez-Vilas (Ed.) Communicating Current Research and Educational Topics and Trends in Applied Microbiology, Formatex. ISBN 978-84-611-9423-0.
- Alzamora, Stella; Tapia, Maria Soledad; López-Malo, Aurelio (2000). Minimally Processed Fruits and Vegetables: Fundamental aspects and applications. Springer. p. 266. ISBN 978-0-8342-1672-3.
- Assessment of technologies for determining cancer risks from the environment. Darby, Pennsylvania, USA: DIANE publishing. 1981. p. 177. ISBN 1-4289-2437-X.
- "The Use of Irradiation for Post-Harvest and Quarantine Commodity Control | Ozone Depletion – Regulatory Programs | U.S. EPA". Wayback.archive.org. Archived from the original on April 21, 2006. Retrieved March 19, 2014.
- "Official Journal of the European Communities. 24 November, 2009. "List of Member States' authorisations of food and food ingredients which may be treated with ionizing radiation."". Retrieved March 19, 2014.
- "Official Journal of the European Communities. 23 October 2002. "COMMISSION DECISION of 23 October 2004 adopting the list of approved facilities in third countries for the irradiation of foods."". Retrieved March 19, 2014.
- "Official Journal of the European Communities. October 13, 2004. "COMMISSION DECISION of October 7, 2004 amending Decision 2002/840/EC adopting the list of approved facilities in third countries for the irradiation of foods."" (PDF). Retrieved March 19, 2014.

5

Effects of Food Contamination

Contaminated food has certain effect on people; the effects discussed in this chapter are foodborne illness, food allergy, diarrhea, nausea and vomiting. Foodborne illness is any illness caused from contaminated food, by bacteria, viruses, or parasites that contaminate food and the symptoms often include vomiting, fever, and may also include diarrhea. The abnormal immune response to food is called food allergy, its symptoms usually include swelling of the tongue, vomiting, trouble breathing or low blood pressure. This chapter helps the reader in understanding the illnesses and problems that can be caused by food contamination.

Foodborne Illness

Foodborne illness (also foodborne disease and colloquially referred to as food poisoning) is any illness resulting from the food spoilage of contaminated food, pathogenic bacteria, viruses, or parasites that contaminate food, as well as chemical or natural toxins such as poisonous mushrooms and various species of beans that have not been boiled for at least 10 minutes.

Symptoms vary depending on the cause, and are described below in this article. A few broad generalizations can be made, e.g.: The incubation period ranges from hours to days, depending on the cause and on how much was consumed. The incubation period tends to cause sufferers to not associate the symptoms with the item consumed, and so to cause sufferers to attribute the symptoms to gastroenteritis for example. Symptoms often include vomiting, fever, and aches, and may include diarrhea. Bouts of vomiting can be repeated with an extended delay in between, because even if infected food was eliminated from the stomach in the first bout, microbes (if applicable) can pass through the stomach into the intestine via cells lining the intestinal walls and begin to multiply. Some types of microbes stay in the intestine, some produce a toxin that is absorbed into the bloodstream, and some can directly invade deeper body tissues.

Causes

Foodborne illness usually arises from improper handling, preparation, or food storage. Good hygiene practices before, during, and after food preparation can reduce the chances of contracting an illness. There is a consensus in the public health community that regular hand-washing is one of the most effective defenses against the spread of foodborne illness. The action of monitoring food to ensure that it will not cause foodborne illness is known as food safety. Foodborne disease can also be caused by a large variety of toxins that affect the environment.

Foodborne illness can also be caused by pesticides or medicines in food and naturally toxic substances such as poisonous mushrooms or reef fish.

Poorly stored food in a refrigerator

Bacteria

Bacteria are a common cause of foodborne illness. In the United Kingdom during 2000, the individual bacteria involved were the following: *Campylobacter jejuni* 77.3%, *Salmonella* 20.9%, *Escherichia coli* O157:H7 1.4%, and all others less than 0.56%. In the past, bacterial infections were thought to be more prevalent because few places had the capability to test for norovirus and no active surveillance was being done for this particular agent. Toxins from bacterial infections are delayed because the bacteria need time to multiply. As a result, symptoms associated with intoxication are usually not seen until 12–72 hours or more after eating contaminated food. However, in some cases, such as Staphylococcal food poisoning, the onset of illness can be as soon as 30 minutes after ingesting contaminated food.

Most common bacterial foodborne pathogens are:

- *Campylobacter jejuni* which can lead to secondary Guillain–Barré syndrome and periodontitis
- *Clostridium perfringens*, the "cafeteria germ"
- *Salmonella* spp. – its *S. typhimurium* infection is caused by consumption of eggs or poultry that are not adequately cooked or by other interactive human-animal pathogens

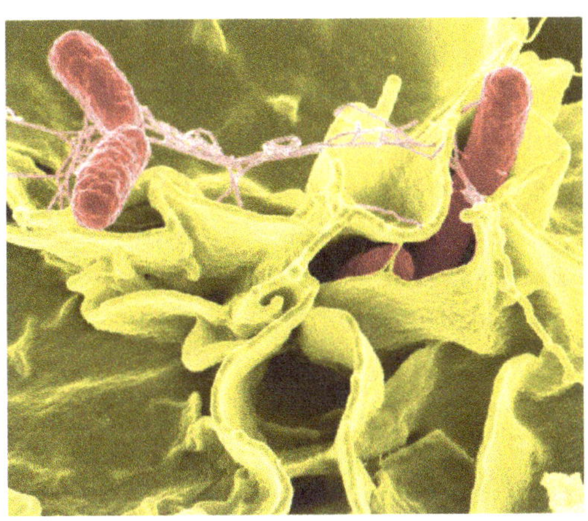

Salmonella

- *Escherichia coli O157:H7* enterohemorrhagic (EHEC) which can cause hemolytic-uremic syndrome

Other common bacterial foodborne pathogens are:

- *Bacillus cereus*
- *Escherichia coli*, other virulence properties, such as enteroinvasive (EIEC), enteropathogenic (EPEC), enterotoxigenic (ETEC), enteroaggregative (EAEC or EAgEC)
- *Listeria monocytogenes*
- *Shigella* spp.
- *Staphylococcus aureus*
- *Staphylococcal enteritis*
- *Streptococcus*
- *Vibrio cholerae*, including O1 and non-O1
- *Vibrio parahaemolyticus*
- *Vibrio vulnificus*
- *Yersinia enterocolitica* and *Yersinia pseudotuberculosis*

Less common bacterial agents:

- *Brucella* spp.
- *Corynebacterium ulcerans*
- *Coxiella burnetii* or Q fever
- *Plesiomonas shigelloides*

Enterotoxins

In addition to disease caused by direct bacterial infection, some foodborne illnesses are caused by enterotoxins (exotoxins targeting the intestines). Enterotoxins can produce illness even when the microbes that produced them have been killed. Symptom appearance varies with the toxin but may be rapid in onset, as in the case of enterotoxins of *Staphylococcus aureus* in which symptoms appear in one to six hours. This causes intense vomiting including or not including diarrhea (resulting in staphylococcal enteritis), and staphylococcal enterotoxins (most commonly staphylococcal enterotoxin A but also including staphylococcal enterotoxin B) are the most commonly reported enterotoxins although cases of poisoning are likely underestimated. It occurs mainly in cooked and processed foods due to competition with other biota in raw foods, and humans are the main cause of contamination as a substantial percentage of humans are persistent carriers of *S. aureus*. The CDC has estimated about 240,000 cases per year in the United States.

- *Clostridium botulinum*
- *Clostridium perfringens*
- *Bacillus cereus*

The rare but potentially deadly disease botulism occurs when the anaerobic bacterium *Clostridium botulinum* grows in improperly canned low-acid foods and produces botulin, a powerful paralytic toxin.

Pseudoalteromonas tetraodonis, certain species of Pseudomonas and Vibrio, and some other bacteria, produce the lethal tetrodotoxin, which is present in the tissues of some living animal species rather than being a product of decomposition.

Emerging Foodborne Pathogens

Many foodborne illnesses remain poorly understood.

- *Aeromonas hydrophila, Aeromonas caviae, Aeromonas sobria*

Preventing Bacterial Food Poisoning

Proper storage and refrigeration of food help in the prevention of food poisoning

Prevention is mainly the role of the state, through the definition of strict rules of hygiene and a public services of veterinary surveying of animal products in the food chain, from farming to the transformation industry and delivery (shops and restaurants). This regulation includes:

- traceability: in a final product, it must be possible to know the origin of the ingredients (originating farm, identification of the harvesting or of the animal) and where and when it was processed; the origin of the illness can thus be tracked and solved (and possibly penalized), and the final products can be removed from the sale if a problem is detected;
- enforcement of hygiene procedures such as HACCP and the "cold chain";
- power of control and of law enforcement of veterinarians.

In August 2006, the United States Food and Drug Administration approved Phage therapy which involves spraying meat with viruses that infect bacteria, and thus preventing infection. This has raised concerns, because without mandatory labelling consumers would not be aware that meat and poultry products have been treated with the spray.

At home, prevention mainly consists of good food safety practices. Many forms of bacterial poisoning can be prevented by cooking it sufficiently, and either eating it quickly or refrigerating it effectively. Many toxins, however, are not destroyed by heat treatment.

Techniques that help prevent food borne illness in the kitchen are hand washing, preventing cross-contamination, proper storage, and maintaining cooking temperatures. In general, freezing or refrigerating prevents virtually all bacteria from growing, and heating food sufficiently kills parasites, viruses, and most bacteria. Bacteria grow most rapidly at the range of temperatures between 40 °F and 140 °F, called the "danger zone". Storing food below or above the "danger zone" can effectively limit the production of toxins. For storing leftovers, the food must be put in shallow containers for quick cooling and must be refrigerated within two hours. When food is reheated, it must reach an internal temperature of 165 °F (74 °C) or until hot or steaming to kill bacteria.

Mycotoxins and Alimentary Mycotoxicoses

The term alimentary mycotoxicoses refers to the effect of poisoning by Mycotoxins (The term 'mycotoxin' is usually reserved for the toxic chemical products produced by fungi that readily colonize crops) through food consumption. Mycotoxins sometimes have important effects on human and animal health. For example, an outbreak which occurred in the UK in 1960 caused the death of 100,000 turkeys which had consumed aflatoxin-contaminated peanut meal. In the USSR in World War II, 5,000 people died due to Alimentary Toxic Aleukia (ALA). The common foodborne Mycotoxins include:

- Aflatoxins – originated from Aspergillus parasiticus and Aspergillus flavus. They are frequently found in tree nuts, peanuts, maize, sorghum and other oilseeds, including corn and cottonseeds. The pronounced forms of Aflatoxins are those of B1, B2, G1, and G2, amongst which Aflatoxin B1 predominantly targets the liver, which will result in necrosis, cirrhosis, and carcinoma. In the US, the acceptable level of total aflatoxins in foods is less than 20 μg/kg, except for Aflatoxin M1 in milk, which should be less than 0.5 μg/kg. The official document can be found at FDA's website.
- Altertoxins – are those of Alternariol (AOH), Alternariol methyl ether (AME), Altenuene (ALT), Altertoxin-1 (ATX-1), Tenuazonic acid (TeA) and Radicinin (RAD), originated from Alternaria spp. Some of the toxins can be present in sorghum, ragi, wheat and tomatoes. Some research has shown that the toxins can be easily cross-contaminated between grain commodities, suggesting that manufacturing and storage of grain commodities is a critical practice.
- Citrinin
- Citreoviridin
- Cyclopiazonic acid
- Cytochalasins
- Ergot alkaloids / Ergopeptine alkaloids – Ergotamine

- Fumonisins – Crop corn can be easily contaminated by the fungi Fusarium moniliforme, and its Fumonisin B1 will cause Leukoencephalomalacia (LEM) in horses, Pulmonary edema syndrome (PES) in pigs, liver cancer in rats and Esophageal cancer in humans. For human and animal health, both the FDA and the EC have regulated the content levels of toxins in food and animal feed.
- Fusaric acid
- Fusarochromanone
- Kojic acid
- Lolitrem alkaloids
- Moniliformin
- 3-Nitropropionic acid
- Nivalenol
- Ochratoxins – In Australia, The Limit of Reporting (LOR) level for Ochratoxin A (OTA) analyses in 20th Australian Total Diet Survey was 1 μg/kg, whereas the EC restricts the content of OTA to 5 μg/kg in cereal commodities, 3 μg/kg in processed products and 10 μg/kg in dried vine fruits.
- Oosporeine
- Patulin – Currently, this toxin has been advisably regulated on fruit products. The EC and the FDA have limited it to under 50 μg/kg for fruit juice and fruit nectar, while limits of 25 μg/kg for solid-contained fruit products and 10 μg/kg for baby foods were specified by the EC.
- Phomopsins
- Sporidesmin A
- Sterigmatocystin
- Tremorgenic mycotoxins – Five of them have been reported to be associated with molds found in fermented meats. These are Fumitremorgen B, Paxilline, Penitrem A, Verrucosidin, and Verruculogen.
- Trichothecenes – sourced from Cephalosporium, Fusarium, Myrothecium, Stachybotrys and Trichoderma. The toxins are usually found in molded maize, wheat, corn, peanuts and rice, or animal feed of hay and straw. Four trichothecenes, T-2 toxin, HT-2 toxin, diacetoxyscirpenol (DAS) and deoxynivalenol (DON) have been most commonly encountered by humans and animals. The consequences of oral intake of, or dermal exposure to, the toxins will result in Alimentary toxic aleukia, neutropenia, aplastic anemia, thrombocytopenia and/or skin irritation. In 1993, the FDA issued a document for the content limits of DON in food and animal feed at an advisory level. In 2003, US published a patent that is very promising for farmers to produce a trichothecene-resistant crop.
- Zearalenone

- Zearalenols

Viruses

Viral infections make up perhaps one third of cases of food poisoning in developed countries. In the US, more than 50% of cases are viral and noroviruses are the most common foodborne illness, causing 57% of outbreaks in 2004. Foodborne viral infection are usually of intermediate (1–3 days) incubation period, causing illnesses which are self-limited in otherwise healthy individuals; they are similar to the bacterial forms described above.

- Enterovirus
- Hepatitis A is distinguished from other viral causes by its prolonged (2–6 week) incubation period and its ability to spread beyond the stomach and intestines into the liver. It often results in jaundice, or yellowing of the skin, but rarely leads to chronic liver dysfunction. The virus has been found to cause infection due to the consumption of fresh-cut produce which has fecal contamination.
- Hepatitis E
- Norovirus
- Rotavirus

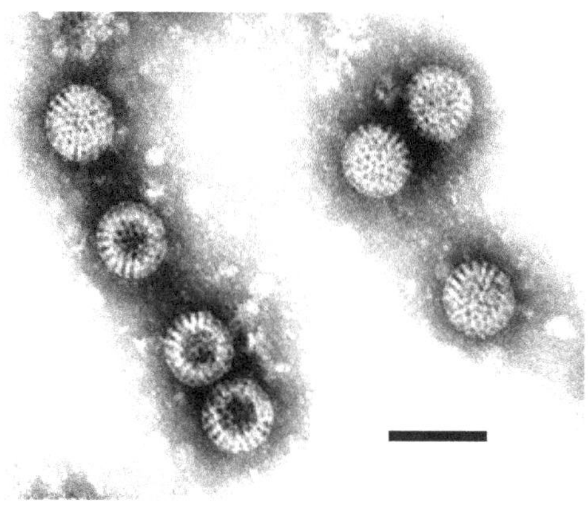

Rotavirus

Parasites

Most foodborne parasites are zoonoses.

- Platyhelminthes:
    - *Diphyllobothrium* sp.
    - *Nanophyetus* sp.
    - *Taenia saginata*
    - *Taenia solium*
    - *Fasciola hepatica*

- Nematode:
  - *Anisakis* sp.
  - *Ascaris lumbricoides*
  - *Eustrongylides* sp.
  - *Trichinella spiralis*
  - *Trichuris trichiura*
- Protozoa:
  - *Acanthamoeba* and other free-living amoebae
  - *Cryptosporidium parvum*
  - *Cyclospora cayetanensis*
  - *Entamoeba histolytica*
  - *Giardia lamblia*

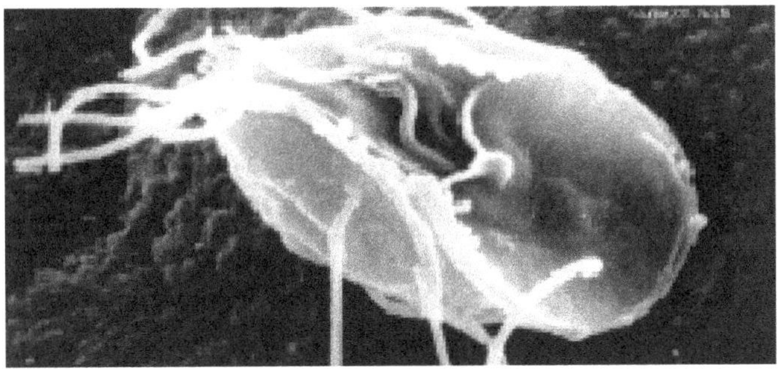

Giardia lamblia

  - *Sarcocystis hominis*
  - *Sarcocystis suihominis*
  - *Toxoplasma gondii*

Natural Toxins

Several foods can naturally contain toxins, many of which are not produced by bacteria. Plants in particular may be toxic; animals which are naturally poisonous to eat are rare. In evolutionary terms, animals can escape being eaten by fleeing; plants can use only passive defenses such as poisons and distasteful substances, for example capsaicin in chili peppers and pungent sulfur compounds in garlic and onions. Most animal poisons are not synthesised by the animal, but acquired by eating poisonous plants to which the animal is immune, or by bacterial action.

- Alkaloids
- Ciguatera poisoning
- Grayanotoxin (honey intoxication)

- Mushroom toxins
- Phytohaemagglutinin (red kidney bean poisoning; destroyed by boiling)
- Pyrrolizidine alkaloids
- Shellfish toxin, including paralytic shellfish poisoning, diarrhetic shellfish poisoning, neurotoxic shellfish poisoning, amnesic shellfish poisoning and ciguatera fish poisoning
- Scombrotoxin
- Tetrodotoxin (fugu fish poisoning)

Some plants contain substances which are toxic in large doses, but have therapeutic properties in appropriate dosages.

- Foxglove contains cardiac glycosides.
- Poisonous hemlock (conium) has medicinal uses.

Other Pathogenic Agents

- Prions, resulting in Creutzfeldt–Jakob disease

"Ptomaine Poisoning"

In 1883, the Italian professor Salmi, of Bologna, introduced the generic name *ptomaine* (from Greek *ptōma*, "fall, fallen body, corpse") for alkaloids found in decaying animal and vegetable matter, especially (as reflected in their names) putrescine and cadaverine. The 1892 *Merck's Bulletin* stated, "We name such products of bacterial origin ptomaines; and the special alkaloid produced by the comma bacillus is variously named Cadaverine, Putrescine, etc." While *The Lancet* stated, "The chemical ferments produced in the system, the...ptomaines which may exercise so disastrous an influence." It is now known that the "disastrous...influence" is due to the direct action of bacteria and only slightly to the alkaloids. Thus, the use of the phrase "ptomaine poisoning" is now obsolete.

Tainted potato salad sickening hundreds at a Communist political convention in Massillon, Ohio, and aboard a Washington DC cruise boat in separate incidents during a single week in 1932 drew national attention to the dangers of so-called "ptomaine poisoning" in the pages of the American news weekly, *Time*. Another newspaper article from 1944 told of more than 150 persons being hospitalized in Chicago with ptomaine poisoning apparently from rice pudding served by a chain of restaurants.

Mechanism

Incubation Period

The delay between consumption of a contaminated food and appearance of the first symptoms of illness is called the incubation period. This ranges from hours to days (and rarely months or even years, such as in the case of listeriosis or bovine spongiform encephalopathy), depending on the agent, and on how much was consumed. If symptoms occur within one to six hours after eating the food, it suggests that it is caused by a bacterial toxin or a chemical rather than live bacteria.

The long incubation period of many foodborne illnesses tends to cause sufferers to attribute their symptoms to gastroenteritis.

During the incubation period, microbes pass through the stomach into the intestine, attach to the cells lining the intestinal walls, and begin to multiply there. Some types of microbes stay in the intestine, some produce a toxin that is absorbed into the bloodstream, and some can directly invade the deeper body tissues. The symptoms produced depend on the type of microbe.

Infectious Dose

The infectious dose is the amount of agent that must be consumed to give rise to symptoms of foodborne illness, and varies according to the agent and the consumer's age and overall health. In the case of *Salmonella* a relatively large inoculum of 1 million to 1 billion organisms is necessary to produce symptoms in healthy human volunteers, as *Salmonellae* are very sensitive to acid. An unusually high stomach pH level (low acidity) greatly reduces the number of bacteria required to cause symptoms by a factor of between 10 and 100.

Epidemiology

United States

In the United States, using FoodNet data from 2000–2007, the CDCP estimated there were 47.8 million foodborne illnesses per year (16,000 cases for 100,000 inhabitants) with 9.4 million of these caused by 31 known identified pathogens.

- 127,839 were hospitalized (43 per 100,000 inhabitants).
- 3,037 people died (1.0 per 100,000 inhabitants).

Causes of foodborne illness in U. S.			
	Cause	**Annual cases**	**Rate (per 100,000 inhabitants)**
1.	Norovirus	5,461,731 cases	X
2.	Salmonella	1,027,561 cases	X
3.	Clostridium perfringens	965,958 cases	X
4.	Campylobacter	845,024 cases	X

Causes of death by foodborne illness in U. S.			
	Cause	**Annual deaths**	**Rate (per 100,000 inhabitants)**
1.	Salmonella	378 cases	0.126
2.	Toxoplasma gondii	327 cases	0.109
3.	Listeria	255 cases	0.085
4.	Norovirus	149 cases	0.050

France

This data pertains to reported medical cases of 23 specific pathogens, as opposed to total population estimates of all food-borne illness for the United States.

In France, for 750,000 cases (1210 per 100,000 inhabitants):

- 70,000 people consulted in the emergency department of an hospital (113 per 100,000 inhabitants.);
- 113,000 people were hospitalized (182 per 100,000 inhabitants);
- 460 people died (0.75 per 100,000 inhabitants).

Causes of foodborne illness in France			
	Cause	**Annual hospitalizations**	**Rate (per 100,000 inhabitants)**
1.	Salmonella	~8,000 cases	13
2.	Campylobacter	~3,000 cases	4.8
3.	Parasites incl. Toxoplasma	~500 cases ~400 cases	0.8 0.65
4.	Listeria	~300 cases	0.5
5.	Hepatitis A	~60 cases	0.1

Causes of death by foodborne illness in France			
	Cause	**Annual**	**Rate (per 100,000 inhabitants)**
1.	Salmonella	~300 cases	0.5
2.	Listeria	~80 cases	0.13
3.	Parasites	~37 cases	0.06 (95% due to toxoplasma)
4.	Campylobacter	~15 cases	0.02
5.	Hepatitis A	~2 cases	0.003

Australia

A study by the Australian National University, published in November 2014, found in 2010 that there were an estimated 4.1 million cases of foodborne gastroenteritis acquired in Australia on average each year, along with 5,140 cases of non-gastrointestinal illness. The study was funded by the Australian Department of Health, Food Standards Australia New Zealand and the NSW Food Authority.

The main causes were Norovirus, pathogenic Escherichia coli, Campylobacter spp. and non-typhoidal Salmonella spp., although the causes of approximately 80% of illnesses were unknown. Approximately 25% (90% CrI: 13%–42%) of the 15.9 million episodes of gastroenteritis that occur in Australia were estimated to be transmitted by contaminated food. This equates to an average of approximately one episode of foodborne gastroenteritis every five years per person. Data on the number of hospitalisations and deaths represent the occurrence of serious foodborne illness. Including gastroenteritis, non-gastroenteritis and sequelae, there were an estimated annual 31,920 (90% CrI: 29,500–35,500) hospitalisations due to foodborne illness and 86 (90% CrI: 70–105) deaths due to foodborne illness circa 2010. This study concludes that these rates are similar to recent estimates in the US and Canada.

A main aim of this study was to compare if foodborne illness incidence had increased over time. In this study, similar methods of assessment were applied to data from circa 2000, which showed that the rate of foodborne gastroenteritis had not changed significantly over time. Two key estimates were the total number of gastroenteritis episodes each year, and the proportion considered foodborne. In circa 2010, it was estimated that 25% of all episodes of gastroenteritis were foodborne. By applying this proportion of episodes due to food to the incidence of gastroenteritis circa 2000, there were an estimated 4.3 million (90% CrI: 2.2–7.3 million) episodes of foodborne gastroenteritis circa 2000, although credible intervals overlap with 2010. Taking into account changes in population size, applying these equivalent methods suggests a 17% decrease in the rate of foodborne gastroenteritis between 2000 and 2010, with considerable overlap of the 90% credible intervals.

This study replaces a previous estimate of 5.4 million cases of food-borne illness in Australia every year, causing:

- 18,000 hospitalizations
- 120 deaths (0.5 deaths per 100,000 inhabitants)
- 2.1 million lost days off work
- 1.2 million doctor consultations
- 300,000 prescriptions for antibiotics.

Most foodborne disease outbreaks in Australia have been linked to raw or minimally cooked eggs or poultry. The Australian Food Safety Information Council estimates that one third of cases of food poisoning occur in the home

Comparison Between Countries

Country	Annual deaths per 100,000 inhabitants	Annual hospitalization per 100,000 inhabitants
USA	1.0	43
France	0.75	182
Australia	0.5	82

Outbreaks

The vast majority of reported cases of foodborne illness occur as individual or sporadic cases. The origin of most sporadic cases is undetermined. In the United States, where people eat outside the home frequently, 58% of cases originate from commercial food facilities (2004 FoodNet data). An outbreak is defined as occurring when two or more people experience similar illness after consuming food from a common source.

Often, a combination of events contributes to an outbreak, for example, food might be left at room temperature for many hours, allowing bacteria to multiply which is compounded by inadequate cooking which results in a failure to kill the dangerously elevated bacterial levels.

Outbreaks are usually identified when those affected know each other. However, more and more, outbreaks are identified by public health staff from unexpected increases in laboratory results for certain strains of bacteria. Outbreak detection and investigation in the United States is primarily handled by local health jurisdictions and is inconsistent from district to district. It is estimated that 1–2% of outbreaks are detected.

Society and Culture

United Kingdom

In postwar Aberdeen (1964) a large-scale (>400 cases) outbreak of typhoid occurred, caused by contaminated corned beef which had been imported from Argentina. The corned beef was placed in cans and because the cooling plant had failed, cold river water from the Plate estuary was used to cool the cans. One of the cans had a defect and the meat inside was contaminated. This meat was then sliced using a meat slicer in a shop in Aberdeen, and a lack of cleaning the machinery led to spreading the contamination to other meats cut in the slicer. These meats were then eaten by the people of Aberdeen who then became ill.

Serious outbreaks of foodborne illness since the 1970s prompted key changes in UK food safety law. These included the death of 19 patients in the Stanley Royd Hospital outbreak and the bovine spongiform encephalopathy (BSE, mad cow disease) outbreak identified in the 1980s. The death of 17 people in the 1996 Wishaw outbreak of *E. coli* O157 was a precursor to the establishment of the Food Standards Agency which, according to Tony Blair in the 1998 white paper *A Force for Change* Cm 3830, "would be powerful, open and dedicated to the interests of consumers".

In May 2015, for the second year running, England's Food Standards Agency devoted its annual Food Safety Week to – "The Chicken Challenge". The focus was on the handling of raw chicken in the home and in catering facilities in a drive to reduce the worryingly high levels of food poisoning from the *campylobacter* bacterium. Anne Hardy argues that widespread public education of food hygiene can be useful, particularly through media (T.V cookery programmes) and advertisement. She points to the examples set by Scandinavian societies.

United States

In 2001, the Center for Science in the Public Interest petitioned the United States Department of Agriculture to require meat packers to remove spinal cords before processing cattle carcasses for human consumption, a measure designed to lessen the risk of infection by variant Creutzfeldt-Jakob disease. The petition was supported by the American Public Health Association, the Consumer Federation of America, the Government Accountability Project, the National Consumers League, and Safe Tables Our Priority. This was opposed by the National Cattlemen's Beef Association, the National Renderers Association, the National Meat Association, the Pork Producers Council, sheep raisers, milk producers, the Turkey Federation, and eight other organizations from the animal-derived food industry.

None of the US Department of Health and Human Services targets regarding incidence of foodborne infections were reached in 2007.

Organizations

The World Health Organization Department of Food Safety and Zoonoses (FOS) provides scientific advice for organizations and the public on issues concerning the safety of food. Its mission is to lower the burden of foodborne disease, thereby strengthening the health security and sustainable development of Member States. Foodborne and waterborne diarrhoeal diseases kill an estimated 2.2 million people annually, most of whom are children. WHO works closely with the Food and Agriculture Organization of the United Nations (FAO) to address food safety issues along the entire food production chain—from production to consumption—using new methods of risk analysis. These methods provide efficient, science-based tools to improve food safety, thereby benefiting both public health and economic development.

International Food Safety Authorities Network (INFOSAN)

The International Food Safety Authorities Network (INFOSAN) is a joint program of the WHO and FAO. INFOSAN has been connecting national authorities from around the globe since 2004, with the goal of preventing the international spread of contaminated food and foodborne disease and strengthening food safety systems globally. This is done by:

1. Promoting the rapid exchange of information during food safety events;
2. Sharing information on important food safety issues of global interest;
3. Promoting partnership and collaboration between countries; and
4. Helping countries strengthen their capacity to manage food safety risks.

Membership to INFOSAN is voluntary, but is restricted to representatives from national and regional government authorities and requires an official letter of designation. INFOSAN seeks to reflect the multidisciplinary nature of food safety and promote intersectoral collaboration by requesting the designation of Focal Points in each of the respective national authorities with a stake in food safety, and a single Emergency Contact Point in the national authority with the responsibility for coordinating national food safety emergencies; countries choosing to be members of INFOSAN are committed to sharing information between their respective food safety authorities and other INFOSAN members. The operational definition of a food safety authority includes those authorities involved in: food policy; risk assessment; food control and management; food inspection services; foodborne disease surveillance and response; laboratory services for monitoring and surveillance of foods and foodborne diseases; and food safety information, education and communication across the farm-to-table continuum.

Regulatory Steps

Food may be contaminated during all stages of food production and retailing. In order to prevent viral contamination, regulatory authorities in Europe have enacted several measures:

- European Commission Regulation (EC) No 2073/2005 of November 15, 2005
- European Committee for Standardization (CEN): Standard method for the detection of norovirus and hepatitis A virus in food products (shellfish, fruits and vegetables, surfaces and bottled water)
- CODEX Committee on Food Hygiene (CCFH): Guideline for the application of general principles of food hygiene for the control of viruses in food

Food Allergy

A food allergy is an abnormal immune response to food. The signs and symptoms may range from mild to severe. They may include itchiness, swelling of the tongue, vomiting, diarrhea, hives, trouble breathing, or low blood pressure. This typically occurs within minutes to several hours of exposure. When the symptoms are severe, it is known as anaphylaxis. Food intolerance and food poisoning are separate conditions.

Common foods involved include cow's milk, peanuts, eggs, shellfish, tree nuts, wheat, rice, and fruit. The common allergies in a region vary depending on the country. Risk factors include a family history of allergies, vitamin D deficiency, obesity, and high levels of cleanliness. Allergies occur when immunoglobulin E (IgE), part of the body's immune system, binds to food molecules. It is usually a protein in the food that is the problem. This triggers the release of inflammatory chemicals such as histamine. Diagnosis is usually based on a medical history, elimination diet, skin prick test, blood tests for food-specific IgE antibodies, or oral food challenge.

Early exposure to potential allergens may be protective. Management primarily involves avoiding the food in question and having a plan if exposure occurs. This plan may include giving adrenaline (epinephrine) and wearing medical alert jewelry. The benefits of allergen immunotherapy for food allergies is unclear and thus not recommended as of 2015. Some types of food allergies among children resolve with age including that to milk, eggs, and soy; while others such as to nuts and shellfish typically do not.

In the developed world about 4% to 8% of people have at least one food allergy. They are more common in children than adults and appear to be increasing in frequency. Male children appear to be more commonly affected than females. Some allergies more commonly develop early in life while others typically develop in later life. In developed countries, a large proportion of people believe they have food allergies when they actually do not have them.

Signs and Symptoms

Food allergies usually have a fast onset (from seconds to one hour) and may include:

- Rash
- Hives
- Itching of mouth, lips, tongue, throat, eyes, skin, or other areas
- Swelling (angioedema) of lips, tongue, eyelids, or the whole face
- Difficulty swallowing
- Runny or congested nose
- Hoarse voice
- Wheezing and/or shortness of breath
- Diarrhea, abdominal pain and/or stomach cramps

- Lightheadedness
- Fainting
- Nausea
- Vomiting

In some cases, however, onset of symptoms may be delayed for hours.

Symptoms of allergies vary from person to person. The amount of food needed to trigger a reaction also varies from person to person.

Serious danger regarding allergies can begin when the respiratory tract or blood circulation is affected. The former can be indicated through wheezing and cyanosis. Poor blood circulation leads to a weak pulse, pale skin and fainting.

A severe case of an allergic reaction, caused by symptoms affecting the respiratory tract and blood circulation, is called anaphylaxis. When symptoms are related to a drop in blood pressure, the person is said to be in anaphylactic shock. Anaphylaxis occurs when IgE Antibodies are involved, and areas of the body that are not in direct contact with the food become affected and show symptoms. This occurs because no nutrients are circulated throughout the body, causing the widening of blood vessels. This vasodilation causes blood pressure to decrease, which leads to the loss of consciousness. Those with asthma or an allergy to peanuts, tree nuts, or seafood are at greater risk for anaphylaxis.

Cause

One of the most common food allergies is a sensitivity to peanuts, a member of the bean family. Peanut allergies may be severe, but children with peanut allergies sometimes outgrow them. Tree nuts, including cashews, Brazil nuts, hazelnuts, macadamia nuts, pecans, pistachios, pine nuts, coconuts, and walnuts, are also common allergens. Sufferers may be sensitive to one particular tree nut or to many different tree nuts. Also seeds, including sesame seeds and poppy seeds, contain oils where protein is present, which may elicit an allergic reaction.

Egg allergies affect about one in fifty children but are frequently outgrown by children when they reach age five. Typically the sensitivity is to proteins in the white, rather than the yolk.

Milk, from cows, goats or sheep, is another common food allergen, and many sufferers are also unable to tolerate dairy products such as cheese. A small portion of children with a milk allergy, roughly ten percent, will have a reaction to beef. Beef contains a small amount of protein that is present in cow's milk.

Other foods containing allergenic proteins include soy, wheat, fish, shellfish, fruits, vegetables, maize, spices, synthetic and natural colors, and chemical additives.

Although sensitivity levels vary by country, the most common food allergies are allergies to milk, eggs, peanuts, tree nuts, seafood, shellfish, soy and wheat. These are often referred to as "the big eight." Allergies to seeds — especially sesame — seem to be increasing in many countries. An example an allergy more common to a particular region is that to rice in East Asia where rice forms a large part of the diet.

Balsam of Peru, which is in various foods, is in the "top five" allergens most commonly causing patch test reactions in people referred to dermatology clinics.

Sensitization

An Institute of Medicine report says that food proteins contained in vaccines, such as gelatin, milk or egg can cause sensitization (development of allergy) in vaccine recipients, to those food items.

Atopy

Food allergies develop more easily in people with the atopic syndrome, a very common combination of diseases: allergic rhinitis and conjunctivitis, eczema and asthma. The syndrome has a strong inherited component; a family history of allergic diseases can be indicative of the atopic syndrome.

Cross Reactivity

Some children who are allergic to cow's milk protein also show a cross sensitivity to soy-based products. There are infant formulas in which the milk and soy proteins are degraded so when taken by an infant, their immune system does not recognize the allergen and they can safely consume the product. Hypoallergenic infant formulas can be based on hydrolyzed proteins, which are proteins partially predigested in a less antigenic form. Other formulas, based on free amino acids, are the least antigenic and provide complete nutrition support in severe forms of milk allergy.

People with latex allergy often also develop allergies to bananas, kiwifruit, avocados, and some other foods.

Pathophysiology

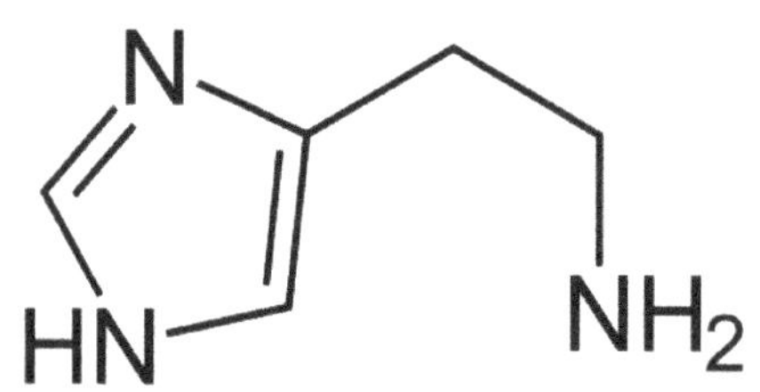

A histamine, the structure shown, is what causes a person to feel itchy during an allergic reaction. A common medication to stop this is an antihistamine, which fights the histamines in the person's system.

A food allergy occurs when the human immune system reacts abnormally to certain foods.

Conditions caused by food allergies are classified into 3 groups according to the mechanism of the allergic response:

1. IgE-mediated (classic) – the most common type, occurs shortly after eating and may involve anaphylaxis.
2. Non-IgE mediated – characterized by an immune response not involving immunoglobulin E; may occur some hours after eating, complicating diagnosis.
3. IgE and/or non-IgE-mediated – a hybrid of the above two types.

Allergic reactions are hyperactive responses of the immune system to generally innocuous substances. When immune cells encounter the allergenic protein, IgE antibodies are produced; this is similar to the immune system's reaction to foreign pathogens. The IgE antibodies identify the allergenic proteins as harmful and initiate the allergic reaction. The harmful proteins are those that do not break down due to the strong bonds of the protein. IgE antibodies bind to a receptor on the surface of the protein, creating a tag, just as a virus or parasite becomes tagged. It is not entirely clear why some proteins do not denature and subsequently trigger allergic reactions and hypersensitivity while others do.

Hypersensitivities are categorized according to the parts of the immune system that are attacked and the amount of time it takes for the response to occur. There are four types of Hypersensitivity reaction: Type 1, Immediate IgE-mediated, Type 2, Cytotoxic, Type 3, Immune complex-mediated, and Type 4, Delayed cell-mediated. The pathophysiology of allergic responses can be divided into two phases. The first is an acute response that occurs immediately after exposure to an allergen. This phase can either subside or progress into a "late phase reaction" which can substantially prolong the symptoms of a response, and result in tissue damage.

Many food allergies are caused by hypersensitivities to particular proteins in different foods. Proteins have unique properties that allow them to become allergens, such as stabilizing forces in the tertiary and quaternary structure which prevent degradation during digestion. Many theoretically allergenic proteins cannot survive the destructive environment of the digestive tract and thus don't trigger hypersensitive reactions.

Acute Response

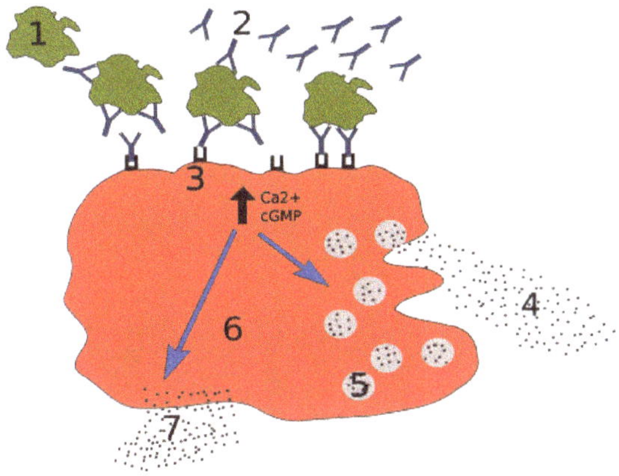

Degranulation process in allergy.
1 — antigen
2 — IgE antibody
3 — FcεRI receptor
4 — preformed mediators (histamine, proteases, chemokines, heparin)
5 — granules
6 — mast cell
7 — newly formed mediators (prostaglandins, leukotrienes, thromboxanes, PAF)

In the early stages of allergy, a type I hypersensitivity reaction against an allergen, encountered for the first time, causes a response in a type of immune cell called a T_H2 lymphocyte, which belongs to a subset of T cells that produce a cytokine called interleukin-4 (IL-4). These T_H2 cells interact with other lymphocytes called B cells, whose role is the production of antibodies. Coupled with signals provided by IL-4, this interaction stimulates the B cell to begin production of a large amount of a particular type of antibody known as IgE. Secreted IgE circulates in the blood and binds to an IgE-specific receptor (a kind of Fc receptor called FcεRI) on the surface of other kinds of immune cells called mast cells and basophils, which are both involved in the acute inflammatory response. The IgE-coated cells, at this stage are sensitized to the allergen.

If later exposure to the same allergen occurs, the allergen can bind to the IgE molecules held on the surface of the mast cells or basophils. Cross-linking of the IgE and Fc receptors occurs when more than one IgE-receptor complex interacts with the same allergenic molecule, and activates the sensitized cell. Activated mast cells and basophils undergo a process called degranulation, during which they release histamine and other inflammatory chemical mediators (cytokines, interleukins, leukotrienes, and prostaglandins) from their granules into the surrounding tissue causing several systemic effects, such as vasodilation, mucous secretion, nerve stimulation and smooth muscle contraction. This results in rhinorrhea, itchiness, dyspnea, and anaphylaxis. Depending on the individual, the allergen, and the mode of introduction, the symptoms can be system-wide (classical anaphylaxis), or localized to particular body systems; asthma is localized to the respiratory system and eczema is localized to the dermis.

Late-phase Response

After the chemical mediators of the acute response subside, late phase responses can often occur. This is due to the migration of other leukocytes such as neutrophils, lymphocytes, eosinophils, and macrophages to the initial site. The reaction is usually seen 2–24 hours after the original reaction. Cytokines from mast cells may also play a role in the persistence of long-term effects. Late phase responses seen in asthma are slightly different from those seen in other allergic responses, although they are still caused by release of mediators from eosinophils, and are still dependent on activity of T_H2 cells.

Diagnosis

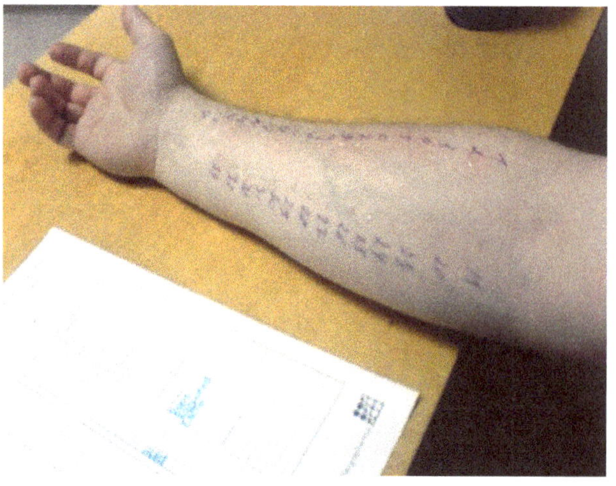

Skin testing on the arm is a common way for detecting an allergy, however, it is not as effective as other tests.

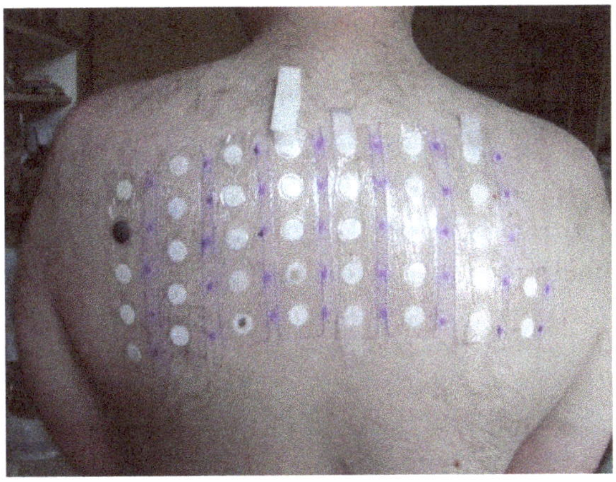

Patch test

Diagnosis is usually based on a medical history, elimination diet, skin prick test, blood tests for food-specific IgE antibodies, or oral food challenge.

- For skin prick tests, a tiny board with protruding needles is used. The allergens are placed either on the board or directly on the skin. The board is then placed on the skin, in order to puncture the skin and for the allergens to enter the body. If a hive appears, the person will be considered positive for the allergy. This test only works for IgE antibodies. Allergic reactions caused by other antibodies cannot be detected through skin prick tests.

Skin prick testing is easy to do and results are available in minutes. Different allergists may use different devices for skin prick testing. Some use a "bifurcated needle", which looks like a fork with 2 prongs. Others use a "multi-test", which may look like a small board with several pins sticking out of it. In these tests, a tiny amount of the suspected allergen is put onto the skin or into a testing device, and the device is placed on the skin to prick, or break through, the top layer of skin. This puts a small amount of the allergen under the skin. A hive will form at any spot where the person is allergic. This test generally yields a positive or negative result. It is good for quickly learning if a person is allergic to a particular food or not, because it detects allergic antibodies known as IgE. Skin tests cannot predict if a reaction would occur or what kind of reaction might occur if a person ingests that particular allergen. They can however confirm an allergy in light of a patient's history of reactions to a particular food. Non-IgE mediated allergies cannot be detected by this method.

- Patch testing is a method used to determine if a specific substance causes allergic inflammation of the skin. It tests for delayed food reactions.

- Blood testing is another way to test for allergies; however, it poses the same disadvantage and only detects IgE allergens and does not work for every possible allergen. RAST, RadioAllergoSorbent Test, is used to detect IgE antibodies present to a certain allergen. The score taken from the RAST test is compared to predictive values, taken from a specific type of RAST test. If the score is higher than the predictive values, there is a great chance the allergy is present in the person. One advantage of this test is that it can test many allergens at one time.

Blood tests are another useful diagnostic tool for evaluating IgE-mediated food allergies. For example, the RAST (RadioAllergoSorbent Test) detects the presence of IgE antibodies to a particular allergen. A CAP-RAST test is a specific type of RAST test with greater specificity: it can show the amount of IgE present to each allergen. Researchers have been able to determine "predictive values" for certain foods. These predictive values can be compared to the RAST blood test results. If a persons RAST score is higher than the predictive value for that food, then there is over a 95% chance the person will have an allergic reaction (limited to rash and anaphylaxis reactions) if they ingest that food. Currently, predictive values are available for the following foods: milk, egg, peanut, fish, soy, and wheat. Blood tests allow for hundreds of allergens to be screened from a single sample, and cover food allergies as well as inhalants. However, non-IgE mediated allergies cannot be detected by this method. Other widely promoted tests such as the *antigen leukocyte cellular antibody test* (ALCAT) and the *Food Allergy Profile* are considered unproven methods, the use of which is not advised.

- Food challenges test for allergens other than those caused by IgE allergens. The allergen is given to the person in the form of a pill, so the person can ingest the allergen directly. The person is watched for signs and symptoms. The problem with food challenges is that they must be performed in the hospital under careful watch, due to the possibility of anaphylaxis.

Food challenges, especially double-blind placebo-controlled food challenges (DBPCFC), are the gold standard for diagnosis of food allergies, including most non-IgE mediated reactions. Blind food challenges involve packaging the suspected allergen into a capsule, giving it to the patient, and observing the patient for signs or symptoms of an allergic reaction.

The best method for diagnosing food allergy is to be assessed by an allergist. The allergist will review the patient's history and the symptoms or reactions that have been noted after food ingestion. If the allergist feels the symptoms or reactions are consistent with food allergy, he/she will perform allergy tests.

- Additional diagnostic tools for evaluation of eosinophilic or non-IgE mediated reactions include endoscopy, colonoscopy, and biopsy.

Differential Diagnosis

Important differential diagnoses are:

- Lactose intolerance; this generally develops later in life but can present in young patients in severe cases. This is due to an enzyme deficiency (lactase) and not allergy. It occurs in many non-Western people.
- Celiac disease; this is an autoimmune disorder triggered by gluten proteins such as gliadin (present in wheat, rye and barley). It is a non-IgE mediated food allergy by definition.
- Irritable bowel syndrome (IBS)
- C1 esterase inhibitor deficiency (hereditary angioedema); this rare disease generally causes attacks of angioedema, but can present solely with abdominal pain and occasional diarrhea.

Prevention

Breastfeeding for more than 4 months may prevent atopic dermatitis, cow milk allergy, and wheezing in early childhood. Early exposure to potential allergens may be protective.

In order to avoid an allergic reaction, a strict diet can be followed. It is difficult to determine the amount of allergenic food required to elicit a reaction, so complete avoidance should be attempted. In some cases, hypersensitive reactions can be triggered by exposures to allergens through skin contact, inhalation, kissing, participation in sports, blood transfusions, cosmetics, and alcohol.

Treatment

The mainstay of treatment for food allergy is total avoidance of the foods that have been identified as allergens. An allergen can enter the body by consuming a food containing the allergen, and can also be ingested by touching any surfaces that may have come into contact with the allergen, then touching the eyes or nose. For people who are extremely sensitive, avoidance includes avoiding touching or inhaling the problematic food.

If the food is accidentally ingested and a systemic reaction (anaphylaxis) occurs, then epinephrine should be used. It is possible that a second dose of epinephrine may be required for severe reactions. The person should then be transported to the emergency room, where additional treatment can be given. Other treatments include antihistamines and steroids.

Epinephrine

Epinephrine auto-injectors are portable single-dose epinephrine-dispensing devices used to treat anaphylaxis.

Epinephrine is another name for the hormone adrenaline, which is produced naturally in the body. An epinephrine injection is the first-line treatment for severe allergic reactions (known as anaphylaxis). If administered in a timely manner, epinephrine can reverse the effects of anaphylaxis.

Epinephrine relieves airway swelling and obstruction and improves blood circulation; blood vessels are tightened and heart rate is increased, improving circulation to bodily organs. Epinephrine is available by prescription in an auto-injector which is used in the treatment of anaphylaxis.

Antihistamines

Antihistamines can alleviate some of the milder symptoms of an allergic reaction, but do not treat all symptoms of anaphylaxis. Antihistamines block the action of histamine, which causes blood vessels to dilate and become leaky to plasma proteins. Histamine also causes itchiness by acting on sensory nerve terminals. The most common antihistamine given for food allergies is diphenhydramine.

Steroids

Glucocorticoid steroids are used to calm down the immune system cells that are attacked by the chemicals released during an allergic reaction. This form of treatment in the form of a nasal spray should not be used to treat anaphylaxis, for it only relieves symptoms in the area in which the steroid is in contact. Another reason steroids should not be used to treat anaphylaxis is due to the long amount of time it takes to reduce inflammation and start to work. Steroids can also be taken orally or through injection. By taking a steroid in these manners, every part of the body can be reached and treated, but a long time is usually needed for these to take effect.

Desensitization

The benefits of allergen immunotherapy for food allergies is unclear and thus not recommended as of 2015. A number of desensitization techniques are; however, being studied.

Epidemiology

The most common food allergens include peanuts, milk, eggs, tree nuts, fish, shellfish, soy, and wheat — these foods account for about 90% of all allergic reactions. The most common food allergies in adults are shellfish, peanuts, tree nuts, fish, and egg. The most common food allergies in children are milk, eggs, peanuts, and tree nuts.

Six to eight percent of children under the age of three have food allergies and nearly four percent of adults have food allergies.

For reasons that are not entirely understood, the diagnosis of food allergies has apparently become more common in Western nations in recent times. In the United States, food allergy affects as many as 5% of infants less than three years of age and 3% to 4% of adults. There is a similar prevalence in Canada.

Seventy-five percent of children who have allergies to milk protein are able to tolerate baked-in milk products, i.e., muffins, cookies, cake and hydrolysed formulas.

About 50% of children with allergies to milk, egg, soy, peanuts, tree nuts and wheat will outgrow their allergy by the age of 6. Those that are still allergic by the age of 12 or so have less than an 8% chance of outgrowing the allergy.

Peanut and tree nut allergies are less likely to be outgrown, although evidence now shows that about 20% of those with peanut allergies and 9% of those with tree nut allergies will outgrow them.

In Central Europe, celery allergy is more common. In Japan, allergy to buckwheat flour, used for soba noodles, is more common.

Meat allergy is extremely rare in the general population, but a geographic cluster of people allergic to meat has been observed in Sydney, Australia. There appears to be a possible association between localised reaction to tick bite and the development of meat allergy.

Fruit allergies exist, such as to: citrus fruits, apples, peaches, pears, jackfruit, strawberries, etc.. This is suspected to be associated with ragweed pollen allergy but could be due to other factors. Reactions can be severe.

Corn allergy may also be prevalent in many populations, although it may be difficult to recognize in areas such as the United States and Canada where corn derivatives are common in the food supply.

Protein allergies or intolerance of seeds, nuts, meat, and milk are especially common among children.

United States

In the United States, it is estimated that up to twelve million people have food allergies.

Food allergies cause roughly 30,000 emergency room visits and 100 to 200 deaths per year and the prevalence is rising. Food allergy affects as many as 5% of infants less than three years of age and 3% to 4% of adults.

For example, an estimated 8% of children have at least one food allergy, a projected total of 5.9 million children across the population, according to a 2011 study published in the journal *Pediatrics*, "The Prevalence, Severity, and Distribution of Childhood Food Allergy in the United States."

Although sensitivity levels vary by country, the most common food allergies are allergies to milk, eggs, peanuts, tree nuts, seafood, shellfish, soy and wheat. These are often referred to as "the big eight." They account for over 90% of the food allergies in the United States.

Society and Culture

In response to the risk that certain foods pose to those with food allergies, some countries have responded by instituting labeling laws that require food products to clearly inform consumers if their products contain major allergens or by-products of major allergens. Some countries also require companies to warn customers when food has been prepared around certain allergens that have been known to cause severe reactions.

From 13 December 2014, new legislation – the EU Food Information for Consumers Regulation 1169/2011 – requires food businesses to provide allergy information on food sold unpackaged, in for example catering outlets, deli counters, bakeries and sandwich bars.

There is much discretion on what food allergies are, and many members of society still believe that a food allergy is simply a dislike of a certain food, and do not realize that it is an actual medical condition.

Law

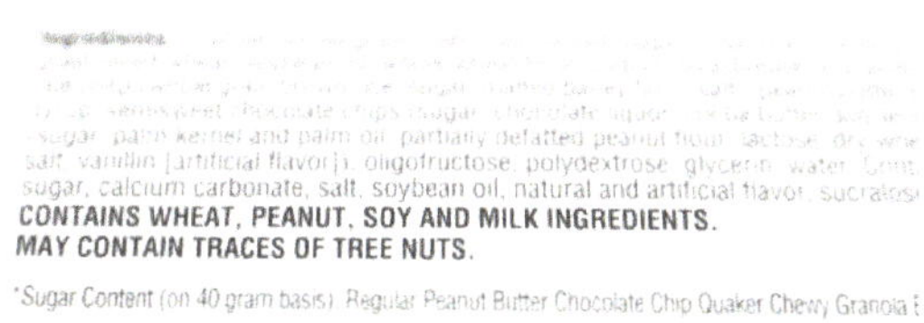

An example of a list of allergens in a food item.

Under the Food Allergen Labeling and Consumer Protection Act of 2004 (Public Law 108-282), companies are required to disclose on the label whether the product contains a major food allergen in clear, plain language. The allergens have to clearly be called out in the ingredient statement. Most companies list allergens in a statement separate from the ingredient statement.

In 2009, Governor Deval Patrick signed into Massachusetts law an Act Relative to Food Allergy Awareness in Restaurants. The allergy awareness act requires food protection managers to view a video about food allergens, a poster identifying the 8 most common food allergens and information about identifying and responding to food allergies posted for food service staff, and customers must be notified of their obligation to inform staff about any food allergies.

On 4 January 2011, President Barack Obama signed into federal law the Food Safety and Modernization Act of 2010 (S510/HR2751, 111th Congress). Section 112 of this Act establishes voluntary food allergy and anaphylaxis management guidelines for public kindergartens, elementary and secondary schools.

Research

Areas of research include anti-IgE antibody (omalizumab) and specific oral tolerance induction (SOTI), which have shown some promise for treatment of certain food allergies.

Gastroenteritis

Gastroenteritis, also known as infectious diarrhea, is inflammation of the gastrointestinal tract that involves the stomach and small intestine. Signs and symptoms include some combination of diarrhea, vomiting, and abdominal pain. Fever, lack of energy, and dehydration may also occur. This typically lasts less than two weeks. It is unrelated to influenza though it has been called the stomach flu.

Gastroenteritis can be due to infections by viruses, bacteria, parasites, and fungus. The most common cause is viruses. In children rotavirus is the most common cause of severe disease. In adults, norovirus and *Campylobacter* are common. Transmission may occur due to eating improperly prepared foods, drinking contaminated water, or through close contact with an individual who is infected. Testing to confirm the diagnosis is typically not needed.

Prevention includes hand washing with soap, drinking clean water, proper disposal of human waste, and breastfeeding babies instead of using formula. The rotavirus vaccine is recommended in children. Treatment involves getting enough fluids. For mild or moderate cases, this can typically be achieved by drinking oral rehydration solution (a combination of water, salts, and sugar). In those who are breast fed, continued breastfeeding is recommended. For more severe cases, intravenous fluids may be needed. Fluids may also be given by a nasogastric tube. Zinc supplementation is recommended in children. Antibiotics are generally not needed.

It is estimated that three to five billion cases of gastroenteritis resulting in 1.4 million deaths occur globally each year. Children and those in the developing world are most commonly affected. As of 2011, in those below age five, there were about 1.7 billion cases resulting in 0.7 million deaths. In the developing world children less than two years of age frequently get six or more infections a year. It is less common in adults, partly due to the development of immunity.

Signs and Symptoms

Gastroenteritis typically involves both diarrhea and vomiting, or less commonly, presents with only one or the other. Abdominal cramping may also be present. Signs and symptoms usually begin 12–72 hours after contracting the infectious agent. If due to a viral agent, the condition usually resolves within one week. Some viral causes may also be associated with fever, fatigue, headache, and muscle pain. If the stool is bloody, the cause is less likely to be viral and more likely to be bacterial. Some bacterial infections may be associated with severe abdominal pain and may persist for several weeks.

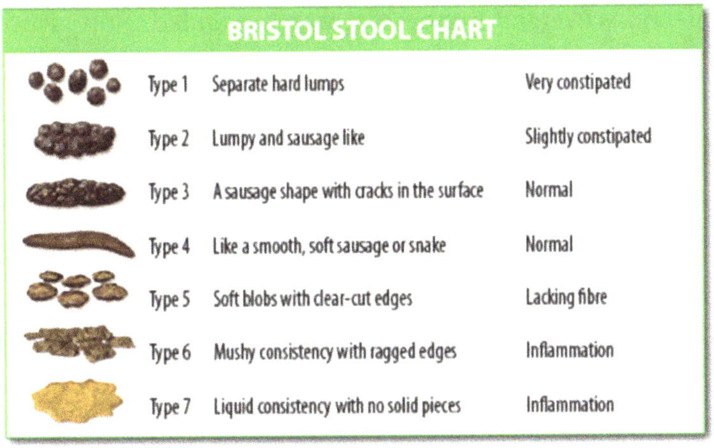

Bristol stool chart

Children infected with rotavirus usually make a full recovery within three to eight days. However, in poor countries treatment for severe infections is often out of reach and persistent diarrhea is common. Dehydration is a common complication of diarrhea, and a child with a significant degree of dehydration may have a prolonged capillary refill, poor skin turgor, and abnormal breathing. Repeat infections are typically seen in areas with poor sanitation, and malnutrition, stunted growth, and long-term cognitive delays can result.

Reactive arthritis occurs in 1% of people following infections with *Campylobacter* species, and Guillain-Barre syndrome occurs in 0.1%. Hemolytic uremic syndrome (HUS) may occur due to infection with Shiga toxin-producing *Escherichia coli* or *Shigella* species, causing low platelet counts, poor kidney function, and low red blood cell count (due to their breakdown). Children are more predisposed to getting HUS than adults. Some viral infections may produce benign infantile seizures.

Cause

Viruses (particularly rotavirus) and the bacteria *Escherichia coli* and *Campylobacter* species are the primary causes of gastroenteritis. There are, however, many other infectious agents that can cause this syndrome. Non-infectious causes are seen on occasion, but they are less likely than a viral or bacterial cause. Risk of infection is higher in children due to their lack of immunity and relatively poor hygiene.

Viral

Rotavirus, norovirus, adenovirus, and astrovirus are known to cause viral gastroenteritis. Rotavirus is the most common cause of gastroenteritis in children, and produces similar rates in both the developed and developing world. Viruses cause about 70% of episodes of infectious diarrhea in the pediatric age group. Rotavirus is a less common cause in adults due to acquired immunity. Norovirus is the cause in about 18% of all cases.

Norovirus is the leading cause of gastroenteritis among adults in America, causing greater than 90% of outbreaks. These localized epidemics typically occur when groups of people spend time in close physical proximity to each other, such as on cruise ships, in hospitals, or in restaurants. People may remain infectious even after their diarrhea has ended. Norovirus is the cause of about 10% of cases in children.

Bacterial

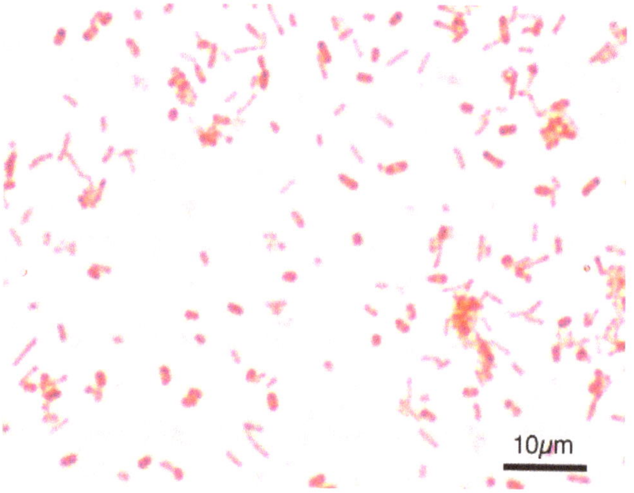

Salmonella enterica serovar Typhimurium (ATCC 14028) as seen with a microscope at 1000 fold magnification and following Gram staining.

In the developed world *Campylobacter jejuni* is the primary cause of bacterial gastroenteritis, with half of these cases associated with exposure to poultry. In children, bacteria are the cause in about 15% of cases, with the most common types being *Escherichia coli*, *Salmonella*, *Shigella*, and *Campylobacter* species. If food becomes contaminated with bacteria and remains at room temperature for a period of several hours, the bacteria multiply and increase the risk of infection in those who consume the food. Some foods commonly associated with illness include raw or undercooked meat, poultry, seafood, and eggs; raw sprouts; unpasteurized milk and soft cheeses; and fruit and vegetable juices. In the developing world, especially sub-Saharan Africa and Asia, cholera is a common cause of gastroenteritis. This infection is usually transmitted by contaminated water or food.

Toxigenic *Clostridium difficile* is an important cause of diarrhea that occurs more often in the elderly. Infants can carry these bacteria without developing symptoms. It is a common cause of diarrhea in those who are hospitalized and is frequently associated with antibiotic use. *Staphylococcus aureus* infectious diarrhea may also occur in those who have used antibiotics. Acute "traveler's diarrhea" is usually a type of bacterial gastroenteritis, while the persistent form is usually parasitic. Acid-suppressing medication appears to increase the risk of significant infection after exposure to a number of organisms, including *Clostridium difficile*, *Salmonella*, and *Campylobacter* species. The risk is greater in those taking proton pump inhibitors than with H2 antagonists.

Parasitic

A number of protozoa can cause gastroenteritis – most commonly *Giardia lamblia* – but *Entamoeba histolytica*, *Cryptosporidium* spp., and other species have also been implicated. As a group, these agents comprise about 10% of cases in children. *Giardia* occurs more commonly in the developing world, but this etiologic agent causes this type of illness to some degree nearly everywhere. It occurs more commonly in persons who have traveled to areas with high prevalence, children who attend day care, men who have sex with men, and following disasters.

Transmission

Transmission may occur via consumption of contaminated water, or when people share personal objects. In places with wet and dry seasons, water quality typically worsens during the wet season, and this correlates with the time of outbreaks. In areas of the world with four seasons, infections are more common in the winter. Bottle-feeding of babies with improperly sanitized bottles is a significant cause on a global scale. Transmission rates are also related to poor hygiene, especially among children, in crowded households, and in those with pre-existing poor nutritional status. After developing tolerance, adults may carry certain organisms without exhibiting signs or symptoms, and thus act as natural reservoirs of contagion. While some agents (such as *Shigella*) only occur in primates, others may occur in a wide variety of animals (such as *Giardia*).

Non-infectious

There are a number of non-infectious causes of inflammation of the gastrointestinal tract. Some of the more common include medications (like NSAIDs), certain foods such as lactose (in those who are intolerant), and gluten (in those with celiac disease). Crohn's disease is also a non-infectious source of (often severe) gastroenteritis. Disease secondary to toxins may also occur. Some food-related conditions associated with nausea, vomiting, and diarrhea include: ciguatera poisoning due to consumption of contaminated predatory fish, scombroid associated with the consumption of certain types of spoiled fish, tetrodotoxin poisoning from the consumption of puffer fish among others, and botulism typically due to improperly preserved food.

In the United States, rates of emergency department use for noninfectious gastroenteritis dropped 30% from 2006 until 2011. Of the twenty most common conditions seen in the emergency department, rates of noninfectious gastroenteritis had the largest decrease in visits in that time period.

Pathophysiology

Gastroenteritis is defined as vomiting or diarrhea due to infection of the small or large bowel. The changes in the small bowel are typically noninflammatory, while the ones in the large bowel are inflammatory. The number of pathogens required to cause an infection varies from as few as one (for *Cryptosporidium*) to as many as 10^8 (for *Vibrio cholerae*).

Diagnosis

Gastroenteritis is typically diagnosed clinically, based on a person's signs and symptoms. Determining the exact cause is usually not needed as it does not alter management of the condition.

However, stool cultures should be performed in those with blood in the stool, those who might have been exposed to food poisoning, and those who have recently traveled to the developing world. Diagnostic testing may also be done for surveillance. As hypoglycemia occurs in approximately 10% of infants and young children, measuring serum glucose in this population is recommended. Electrolytes and kidney function should also be checked when there is a concern about severe dehydration.

Dehydration

A determination of whether or not the person has dehydration is an important part of the assessment, with dehydration typically divided into mild (3–5%), moderate (6–9%), and severe (≥10%) cases. In children, the most accurate signs of moderate or severe dehydration are a prolonged capillary refill, poor skin turgor, and abnormal breathing. Other useful findings (when used in combination) include sunken eyes, decreased activity, a lack of tears, and a dry mouth. A normal urinary output and oral fluid intake is reassuring. Laboratory testing is of little clinical benefit in determining the degree of dehydration. Thus the use of urine testing or ultrasounds is generally not needed.

Differential Diagnosis

Other potential causes of signs and symptoms that mimic those seen in gastroenteritis that need to be ruled out include appendicitis, volvulus, inflammatory bowel disease, urinary tract infections, and diabetes mellitus. Pancreatic insufficiency, short bowel syndrome, Whipple's disease, coeliac disease, and laxative abuse should also be considered. The differential diagnosis can be complicated somewhat if the person exhibits *only* vomiting or diarrhea (rather than both).

Appendicitis may present with vomiting, abdominal pain, and a small amount of diarrhea in up to 33% of cases. This is in contrast to the large amount of diarrhea that is typical of gastroenteritis. Infections of the lungs or urinary tract in children may also cause vomiting or diarrhea. Classical diabetic ketoacidosis (DKA) presents with abdominal pain, nausea, and vomiting, but without diarrhea. One study found that 17% of children with DKA were initially diagnosed as having gastroenteritis.

Prevention

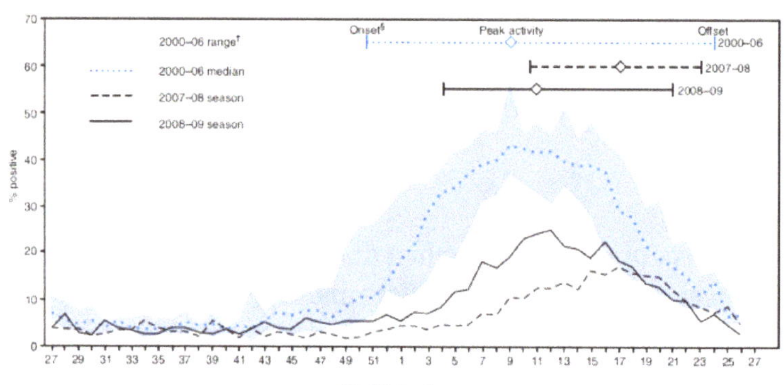

Percentage of rotavirus tests with positive results, by surveillance week, United States, July 2000 – June 2009.

Lifestyle

A supply of easily accessible uncontaminated water and good sanitation practices are important for reducing rates of infection and clinically significant gastroenteritis. Personal measures (such as hand washing with soap) have been found to decrease rates of gastroenteritis in both the developing and developed world by as much as 30%. Alcohol-based gels may also be effective. Breastfeeding is important, especially in places with poor hygiene, as is improvement of hygiene generally. Breast milk reduces both the frequency of infections and their duration. Avoiding contaminated food or drink should also be effective.

Vaccination

Due to both its effectiveness and safety, in 2009 the World Health Organization recommended that the rotavirus vaccine be offered to all children globally. Two commercial rotavirus vaccines exist and several more are in development. In Africa and Asia these vaccines reduced severe disease among infants and countries that have put in place national immunization programs have seen a decline in the rates and severity of disease. This vaccine may also prevent illness in non-vaccinated children by reducing the number of circulating infections. Since 2000, the implementation of a rotavirus vaccination program in the United States has substantially decreased the number of cases of diarrhea by as much as 80 percent. The first dose of vaccine should be given to infants between 6 and 15 weeks of age. The oral cholera vaccine has been found to be 50–60% effective over 2 years.

Management

Gastroenteritis is usually an acute and self-limiting disease that does not require medication. The preferred treatment in those with mild to moderate dehydration is oral rehydration therapy (ORT). Metoclopramide and/or ondansetron, however, may be helpful in some children,[*needs update*] and butylscopolamine is useful in treating abdominal pain.

Rehydration

The primary treatment of gastroenteritis in both children and adults is rehydration. This is preferably achieved by drinking rehydration solution, although intravenous delivery may be required if there is a decreased level of consciousness or if dehydration is severe. Drinking replacement therapy products made with complex carbohydrates (i.e. those made from wheat or rice) may be superior to those based on simple sugars. Drinks especially high in simple sugars, such as soft drinks and fruit juices, are not recommended in children under 5 years of age as they may *increase* diarrhea. Plain water may be used if more specific ORT preparations are unavailable or the person is not willing to drink them. A nasogastric tube can be used in young children to administer fluids if warranted. In those who require intravenous fluids, one to four hours' worth is often sufficient.

Dietary

It is recommended that breast-fed infants continue to be nursed in the usual fashion, and that formula-fed infants continue their formula immediately after rehydration with ORT. Lactose-free or lactose-reduced formulas usually are not necessary. Children should continue their usual diet during episodes of diarrhea with the exception that foods high in simple sugars should be avoided. The BRAT diet (bananas, rice, applesauce, toast and tea) is no longer recommended, as it contains insufficient nutrients and has no benefit over normal feeding. Some probiotics have been shown to be beneficial in reducing both the duration of illness and the frequency of stools. They may also be useful in preventing and treating antibiotic associated diarrhea. Fermented milk products (such as yogurt) are similarly beneficial. Zinc supplementation appears to be effective in both treating and preventing diarrhea among children in the developing world.

Antiemetics

Antiemetic medications may be helpful for treating vomiting in children. Ondansetron has some utility, with a single dose being associated with less need for intravenous fluids, fewer hospitalizations, and decreased vomiting. Metoclopramide might also be helpful. However, the use of ondansetron might possibly be linked to an increased rate of return to hospital in children. The intravenous preparation of ondansetron may be given orally if clinical judgment warrants. Dimenhydrinate, while reducing vomiting, does not appear to have a significant clinical benefit.

Antibiotics

Antibiotics are not usually used for gastroenteritis, although they are sometimes recommended if symptoms are particularly severe or if a susceptible bacterial cause is isolated or suspected. If antibiotics are to be employed, a macrolide (such as azithromycin) is preferred over a fluoroquinolone due to higher rates of resistance to the latter. Pseudomembranous colitis, usually caused by antibiotic use, is managed by discontinuing the causative agent and treating it with either metronidazole or vancomycin. Bacteria and protozoans that are amenable to treatment include *Shigella Salmonella typhi*, and *Giardia* species. In those with *Giardia* species or *Entamoeba histolytica*, tinidazole treatment is recommended and superior to metronidazole. The World Health Organization (WHO) recommends the use of antibiotics in young children who have both bloody diarrhea and fever.

Antimotility Agents

Antimotility medication has a theoretical risk of causing complications, and although clinical experience has shown this to be unlikely, these drugs are discouraged in people with bloody diarrhea or diarrhea that is complicated by fever. Loperamide, an opioid analogue, is commonly used for the symptomatic treatment of diarrhea. Loperamide is not recommended in children, however, as it may cross the immature blood–brain barrier and cause toxicity. Bismuth subsalicylate, an insoluble complex of trivalent bismuth and salicylate, can be used in mild to moderate cases, but salicylate toxicity is theoretically possible.

Epidemiology

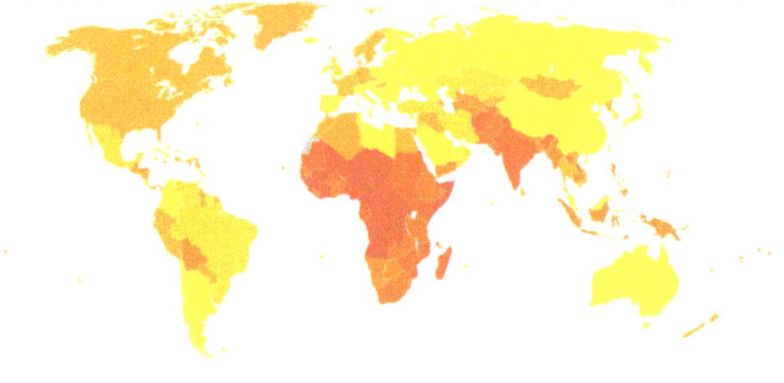

Deaths due to diarrhoeal diseases per million persons in 2012

- 0-2
- 3-10
- 11-18
- 19-30
- 31-46
- 47-80
- 81-221
- 222-450
- 451-606
- 607-1799

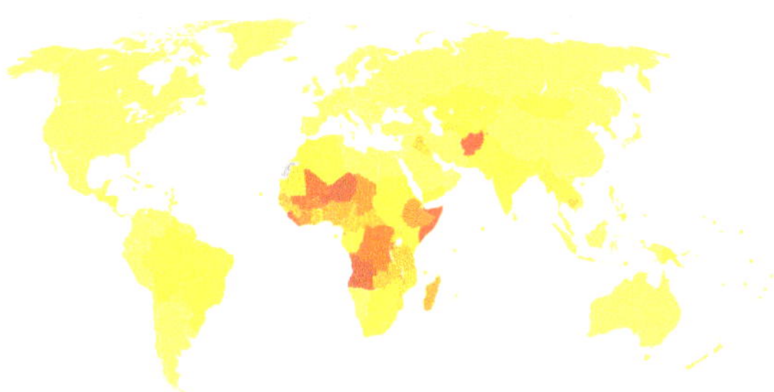

Disability-adjusted life year for diarrhea per 100,000 inhabitants in 2004.

no data	3000–3500
≤less 500	3500–4000
500–1000	4000–4500
1000–1500	4500–5000
1500–2000	5000–6000
2000–2500	≥6000
2500–3000	

It is estimated that three to five billion cases of gastroenteritis resulting in 1.4 million deaths occur globally on an annual basis. Children and those in the developing world are most commonly affected. As of 2011, in those less than five, there were about 1.7 billion cases resulting in 0.7 million deaths, with most of these occurring in the world's poorest nations. More than 450,000 of these fatalities are due to rotavirus in children under 5 years of age. Cholera causes about three to five million cases of disease and kills approximately 100,000 people yearly. In the developing world, children less than two years of age frequently get six or more infections a year that result in significant gastroenteritis. It is less common in adults, partly due to the development of acquired immunity.

In 1980, gastroenteritis from all causes caused 4.6 million deaths in children, with the majority occurring in the developing world. Death rates were reduced significantly (to approximately 1.5 million deaths annually) by the year 2000, largely due to the introduction and widespread use of oral rehydration therapy. In the US, infections causing gastroenteritis are the second most common infection (after the common cold), and they result in between 200 and 375 million cases of acute diarrhea and approximately ten thousand deaths annually, with 150 to 300 of these deaths in children less than five years of age.

History

The first usage of "gastroenteritis" was in 1825. Before this time it was commonly known as typhoid fever or "cholera morbus", among others, or less specifically as "griping of the guts", "surfeit", "flux", "colic", "bowel complaint", or any one of a number of other archaic names for acute diarrhea. Cholera morbus is a historical term that was used to refer to gastroenteritis rather than specifically cholera.

Society and Culture

Gastroenteritis is associated with many colloquial names, including "Montezuma's revenge", "Delhi belly", "la turista", and "back door sprint", among others. It has played a role in many military campaigns and is believed to be the origin of the term "no guts no glory".

Gastroenteritis is the main reason for 3.7 million visits to physicians a year in the United States and 3 million visits in France. In the United States gastroenteritis as a whole is believed to result in costs of 23 billion USD per year with that due to rotavirus alone resulting in estimated costs of 1 billion USD a year.

Research

There are a number of vaccines against gastroenteritis in development. For example, vaccines against *Shigella* and enterotoxigenic *Escherichia coli* (ETEC), two of the leading bacterial causes of gastroenteritis worldwide.

Other Animals

Many of the same agents cause gastroenteritis in cats and dogs as in humans. The most common organisms are *Campylobacter*, *Clostridium difficile*, *Clostridium perfringens*, and *Salmonella*. A large number of toxic plants may also cause symptoms.

Some agents are more specific to a certain species. Transmissible gastroenteritis coronavirus (TGEV) occurs in pigs resulting in vomiting, diarrhea, and dehydration. It is believed to be introduced to pigs by wild birds and there is no specific treatment available. It is not transmissible to humans.

Diarrhea

Diarrhea, also spelled diarrhoea, is the condition of having at least three loose or liquid bowel movements each day. It often lasts for a few days and can result in dehydration due to fluid loss.

Signs of dehydration often begin with loss of the normal stretchiness of the skin and irritable behaviour. This can progress to decreased urination, loss of skin color, a fast heart rate, and a decrease in responsiveness as it becomes more severe. Loose but non-watery stools in babies who are breastfed, however, may be normal.

The most common cause is an infection of the intestines due to either a virus, bacteria, or parasite; a condition known as gastroenteritis. These infections are often acquired from food or water that has been contaminated by stool, or directly from another person who is infected. It may be divided into three types: short duration watery diarrhea, short duration bloody diarrhea, and if it lasts for more than two weeks, persistent diarrhea. The short duration watery diarrhea may be due to an infection by cholera, although this is rare in the developed world. If blood is present it is also known as dysentery. A number of non-infectious causes may also result in diarrhea, including hyperthyroidism, lactose intolerance, inflammatory bowel disease, a number of medications, and irritable bowel syndrome. In most cases, stool cultures are not required to confirm the exact cause.

Prevention of infectious diarrhea is by improved sanitation, clean drinking water, and hand washing with soap. Breastfeeding for at least six months is also recommended as is vaccination against rotavirus. Oral rehydration solution (ORS), which is clean water with modest amounts of salts and sugar, is the treatment of choice. Zinc tablets are also recommended. These treatments have been estimated to have saved 50 million children in the past 25 years. When people have diarrhea it is recommended that they continue to eat healthy food and babies continue to be breastfed. If commercial ORS are not available, homemade solutions may be used. In those with severe dehydration, intravenous fluids may be required. Most cases; however, can be managed well with fluids by mouth. Antibiotics, while rarely used, may be recommended in a few cases such as those who have bloody diarrhea and a high fever, those with severe diarrhea following travelling, and those who grow specific bacteria or parasites in their stool. Loperamide may help decrease the number of bowel movements but is not recommended in those with severe disease.

About 1.7 to 5 billion cases of diarrhea occur per year. It is most common in developing countries, where young children get diarrhea on average three times a year. Total deaths from diarrhea are estimated at 1.26 million in 2013 – down from 2.58 million in 1990. In 2012, it was the second most common cause of deaths in children younger than five (0.76 million or 11%). Frequent episodes of diarrhea are also a common cause of malnutrition and the most common cause in those younger than five years of age. Other long term problems that can result include stunted growth and poor intellectual development.

Definition

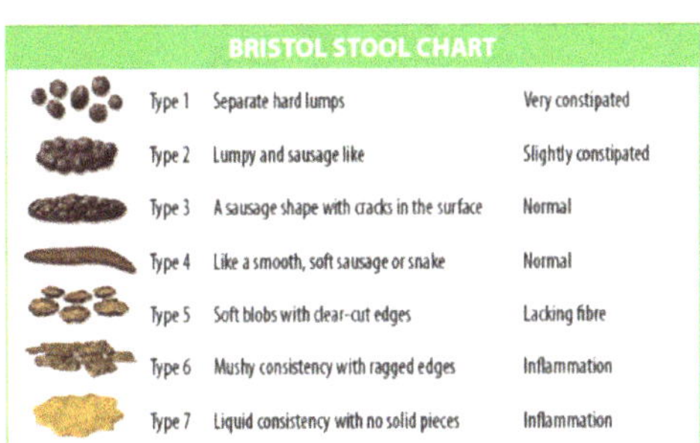

Bristol stool chart

Diarrhea is defined by the World Health Organization as having three or more loose or liquid stools per day, or as having more stools than is normal for that person.

Acute diarrhea is defined as an abnormally frequent discharge of semisolid or fluid fecal matter from the bowel, lasting less than 14 days, by World Gastroenterology Organization.

Secretory

Secretory diarrhea means that there is an increase in the active secretion, or there is an inhibition of absorption. There is little to no structural damage. The most common cause of this type of diarrhea is a cholera toxin that stimulates the secretion of anions, especially chloride ions. Therefore, to maintain a charge balance in the gastrointestinal tract, sodium is carried with it, along with water. In this type of diarrhea intestinal fluid secretion is isotonic with plasma even during fasting. It continues even when there is no oral food intake.

Osmotic

Osmotic diarrhea occurs when too much water is drawn into the bowels. If a person drinks solutions with excessive sugar or excessive salt, these can draw water from the body into the bowel and cause osmotic diarrhea. Osmotic diarrhea can also be the result of maldigestion (e.g., pancreatic disease or Coeliac disease), in which the nutrients are left in the lumen to pull in water. Or it can be caused by osmotic laxatives (which work to alleviate constipation by drawing water into the bowels). In healthy individuals, too much magnesium or vitamin C or undigested lactose can produce osmotic diarrhea and distention of the bowel. A person who has lactose intolerance can have difficulty absorbing lactose after an extraordinarily high intake of dairy products. In persons who have fructose malabsorption, excess fructose intake can also cause diarrhea. High-fructose foods that also have a high glucose content are more absorbable and less likely to cause diarrhea. Sugar alcohols such as sorbitol (often found in sugar-free foods) are difficult for the body to absorb and, in large amounts, may lead to osmotic diarrhea. In most of these cases, osmotic diarrhea stops when the offending agent (e.g. milk, sorbitol) is stopped.

Exudative

Exudative diarrhea occurs with the presence of blood and pus in the stool. This occurs with inflammatory bowel diseases, such as Crohn's disease or ulcerative colitis, and other severe infections such as *E. coli* or other forms of food poisoning.

Inflammatory

Inflammatory diarrhea occurs when there is damage to the mucosal lining or brush border, which leads to a passive loss of protein-rich fluids and a decreased ability to absorb these lost fluids. Features of all three of the other types of diarrhea can be found in this type of diarrhea. It can be caused by bacterial infections, viral infections, parasitic infections, or autoimmune problems such as inflammatory bowel diseases. It can also be caused by tuberculosis, colon cancer, and enteritis.

Dysentery

If there is blood visible in the stools, it is also known as dysentery. The blood is a trace of an invasion of bowel tissue. Dysentery is a symptom of, among others, *Shigella*, *Entamoeba histolytica*, and *Salmonella*.

Health Effects

Diarrheal disease may have a negative impact on both physical fitness and mental development. "Early childhood malnutrition resulting from any cause reduces physical fitness and work productivity in adults," and diarrhea is a primary cause of childhood malnutrition. Further, evidence suggests that diarrheal disease has significant impacts on mental development and health; it has been shown that, even when controlling for helminth infection and early breastfeeding, children who had experienced severe diarrhea had significantly lower scores on a series of tests of intelligence.

Differential Diagnosis

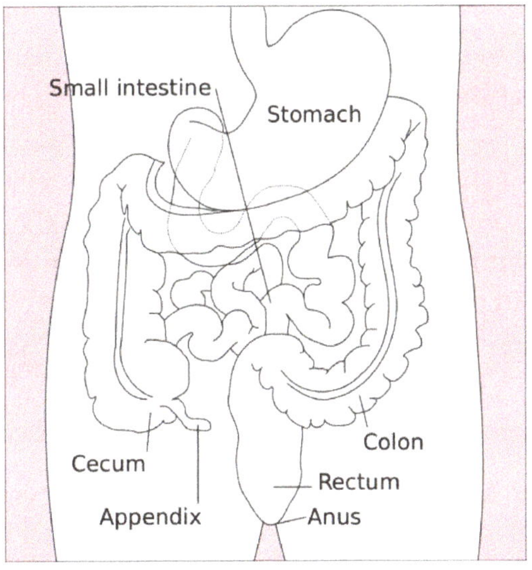

Diagram of the human gastrointestinal tract.

Acute diarrhea is most commonly due to viral gastroenteritis with rotavirus, which accounts for 40% of cases in children under five. (p. 17) In travelers however bacterial infections predominate. Various toxins such as mushroom poisoning and drugs can also cause acute diarrhea.

Chronic diarrhea can be the part of the presentations of a number of chronic medical conditions affecting the intestine. Common causes include ulcerative colitis, Crohn's disease, microscopic colitis, celiac disease, irritable bowel syndrome and bile acid malabsorption.

Infections

There are many causes of infectious diarrhea, which include viruses, bacteria and parasites. Infectious diarrhea is frequently referred to as gastroenteritis. Norovirus is the most common cause of viral diarrhea in adults, but rotavirus is the most common cause in children under five years old. Adenovirus types 40 and 41, and astroviruses cause a significant number of infections.

Campylobacter spp. are a common cause of bacterial diarrhea, but infections by *Salmonella* spp., *Shigella* spp. and some strains of *Escherichia coli* are also a frequent cause.

In the elderly, particularly those who have been treated with antibiotics for unrelated infections, a toxin produced by *Clostridium difficile* often causes severe diarrhea.

Parasites, particularly protozoa (e.g., *Cryptosporidium* spp., *Giardia* spp., *Entamoeba histolytica*, *Blastocystis* spp., *Cyclospora cayetanensis*), are frequently the cause of diarrhea that involves chronic infection. The broad-spectrum antiparasitic agent nitazoxanide has shown efficacy against many diarrhea-causing parasites.

Other infectious agents, such as parasites or bacterial toxins, may exacerbate symptoms. In sanitary living conditions where there is ample food and a supply of clean water, an otherwise healthy person usually recovers from viral infections in a few days. However, for ill or malnourished individuals, diarrhea can lead to severe dehydration and can become life-threatening.

Malabsorption

Malabsorption is the inability to absorb food fully, mostly from disorders in the small bowel, but also due to maldigestion from diseases of the pancreas.

Causes include:

- *enzyme deficiencies or mucosal abnormality*, as in food allergy and food intolerance, e.g. celiac disease (gluten intolerance), lactose intolerance (intolerance to milk sugar, common in non-Europeans), and fructose malabsorption.
- *pernicious anemia*, or impaired bowel function due to the inability to absorb vitamin B_{12},
- *loss of pancreatic secretions*, which may be due to cystic fibrosis or pancreatitis,
- *structural defects*, like short bowel syndrome (surgically removed bowel) and radiation fibrosis, such as usually follows cancer treatment and other drugs, including agents used in chemotherapy; and
- *certain drugs*, like orlistat, which inhibits the absorption of fat.

Inflammatory Bowel Disease

The two overlapping types here are of unknown origin:

- Ulcerative colitis is marked by chronic bloody diarrhea and inflammation mostly affects the distal colon near the rectum.
- Crohn's disease typically affects fairly well demarcated segments of bowel in the colon and often affects the end of the small bowel.

Irritable Bowel Syndrome

Another possible cause of diarrhea is irritable bowel syndrome (IBS), which usually presents with abdominal discomfort relieved by defecation and unusual stool (diarrhea or constipation) for at least 3 days a week over the previous 3 months. Symptoms of diarrhea-predominant IBS can be managed through a combination of dietary changes, soluble fiber supplements, and/or medications such as loperamide or codeine. About 30% of patients with diarrhea-predominant IBS have bile acid malabsorption diagnosed with an abnormal SeHCAT test.

Other Diseases

Diarrhea can be caused by other diseases and conditions, namely:

- Chronic ethanol ingestion
- Ischemic bowel disease: This usually affects older people and can be due to blocked arteries.
- Microscopic colitis, a type of inflammatory bowel disease where changes are only seen on histological examination of colonic biopsies.
- Bile salt malabsorption (primary bile acid diarrhea) where excessive bile acids in the colon produce a secretory diarrhea.
- Hormone-secreting tumors: some hormones (e.g., serotonin) can cause diarrhea if excreted in excess (usually from a tumor).
- Chronic mild diarrhea in infants and toddlers may occur with no obvious cause and with no other ill effects; this condition is called toddler's diarrhea.
- Environmental enteropathy
- Radiation enteropathy following treatment for pelvic and abdominal cancers.

Causes

Sanitation

Poverty often leads to unhygienic living conditions, as in this community in the Indian Himalayas. Such conditions promote contraction of diarrheal diseases, as a result of poor sanitation and hygiene.

Open defecation is a leading cause of infectious diarrhea leading to death.

Poverty is a good indicator of the rate of infectious diarrhea in a population. This association does not stem from poverty itself, but rather from the conditions under which impoverished people live. The absence of certain resources compromises the ability of the poor to defend themselves against infectious diarrhea. “Poverty is associated with poor housing, crowding, dirt floors, lack of access to clean water or to sanitary disposal of fecal waste (sanitation), cohabitation with domestic animals that may carry human pathogens, and a lack of refrigerated storage for food, all of which increase the frequency of diarrhea... Poverty also restricts the ability to provide age-appropriate, nutritionally balanced diets or to modify diets when diarrhea develops so as to mitigate and repair nutrient losses. The impact is exacerbated by the lack of adequate, available, and affordable medical care.”

Water

One of the most common causes of infectious diarrhea, is a lack of clean water. Often, improper fecal disposal leads to contamination of groundwater. This can lead to widespread infection among a population, especially in the absence of water filtration or purification. Human feces contains a variety of potentially harmful human pathogens.

Nutrition

Proper nutrition is important for health and functioning, including the prevention of infectious diarrhea. It is especially important to young children who do not have a fully developed immune system. Zinc deficiency, a condition often found in children in developing countries can, even in mild cases, have a significant impact on the development and proper functioning of the human immune system. Indeed, this relationship between zinc deficiency and reduced immune functioning corresponds with an increased severity of infectious diarrhea. Children who have lowered levels of zinc have a greater number of instances of diarrhea, severe diarrhea, and diarrhea associated with fever. Similarly, vitamin A deficiency can cause an increase in the severity of diarrheal episodes. However, there is some discrepancy when it comes to the impact of vitamin A deficiency on the rate of disease. While some argue that a relationship does not exist between the rate of disease and vitamin A status, others suggest an increase in the rate associated with deficiency. Given that estimates suggest 127 million preschool children worldwide are vitamin A deficient, this population has the potential for increased risk of disease contraction.

Pathophysiology

Evolution

According to two researchers, Nesse and Williams, diarrhea may function as an evolved expulsion defense mechanism. As a result, if it is stopped, there might be a delay in recovery. They cite in support of this argument research published in 1973 that found that treating *Shigella* with the anti-diarrhea drug (Co-phenotrope, Lomotil) caused people to stay feverish twice as long as those not so treated. The researchers indeed themselves observed that: “Lomotil may be contraindicated in shigellosis. Diarrhea may represent a defense mechanism”.

Diagnostic Approach

The following types of diarrhea may indicate further investigation is needed:

- In infants

- Moderate or severe diarrhea in young children
- Associated with blood
- Continues for more than two days
- Associated non-cramping abdominal pain, fever, weight loss, etc.
- In travelers
- In food handlers, because of the potential to infect others;
- In institutions such as hospitals, child care centers, or geriatric and convalescent homes.

A severity score is used to aid diagnosis in children.

Prevention

Sanitation

Numerous studies have shown that improvements in drinking water and sanitation (WASH) lead to decreased risks of diarrhoea. Such improvements might include for example use of water filters, provision of high-quality piped water and sewer connections.

In institutions, communities, and households, interventions that promote hand washing with soap lead to significant reductions in the incidence of diarrhea. The same applies to preventing open defecation at a community-wide level and providing access to improved sanitation. This includes use of toilets and implementation of the entire sanitation chain connected to the toilets (collection, transport, disposal or reuse of human excreta).

Hand Washing

Basic sanitation techniques can have a profound effect on the transmission of diarrheal disease. The implementation of hand washing using soap and water, for example, has been experimentally shown to reduce the incidence of disease by approximately 42–48%. Hand washing in developing countries, however, is compromised by poverty as acknowledged by the CDC: "Handwashing is integral to disease prevention in all parts of the world; however, access to soap and water is limited in a number of less developed countries. This lack of access is one of many challenges to proper hygiene in less developed countries." Solutions to this barrier require the implementation of educational programs that encourage sanitary behaviours.

Water

Given that water contamination is a major means of transmitting diarrheal disease, efforts to provide clean water supply and improved sanitation have the potential to dramatically cut the rate of disease incidence. In fact, it has been proposed that we might expect an 88% reduction in child mortality resulting from diarrheal disease as a result of improved water sanitation and hygiene. Similarly, a meta-analysis of numerous studies on improving water supply and sanitation shows a 22–27% reduction in disease incidence, and a 21–30% reduction in mortality rate associated with diarrheal disease.

Chlorine treatment of water, for example, has been shown to reduce both the risk of diarrheal disease, and of contamination of stored water with diarrheal pathogens.

Vaccination

Immunization against the pathogens that cause diarrheal disease is a viable prevention strategy, however it does require targeting certain pathogens for vaccination. In the case of Rotavirus, which was responsible for around 6% of diarrheal episodes and 20% of diarrheal disease deaths in the children of developing countries, use of a Rotavirus vaccine in trials in 1985 yielded a slight (2-3%) decrease in total diarrheal disease incidence, while reducing overall mortality by 6-10%. Similarly, a Cholera vaccine showed a strong reduction in morbidity and mortality, though the overall impact of vaccination was minimal as Cholera is not one of the major causative pathogens of diarrheal disease. Since this time, more effective vaccines have been developed that have the potential to save many thousands of lives in developing nations, while reducing the overall cost of treatment, and the costs to society.

A rotavirus vaccine decrease the rates of diarrhea in a population. New vaccines against rotavirus, *Shigella*, Enterotoxigenic Escherichia coli (ETEC), and cholera are under development, as well as other causes of infectious diarrhea.

Nutrition

Dietary deficiencies in developing countries can be combated by promoting better eating practices. Supplementation with vitamin A and/or zinc. Zinc supplementation proved successful showing a significant decrease in the incidence of diarrheal disease compared to a control group. The majority of the literature suggests that vitamin A supplementation is advantageous in reducing disease incidence. Development of a supplementation strategy should take into consideration the fact that vitamin A supplementation was less effective in reducing diarrhea incidence when compared to vitamin A and zinc supplementation, and that the latter strategy was estimated to be significantly more cost effective.

Breastfeeding

Breastfeeding practices have been shown to have a dramatic effect on the incidence of diarrheal disease in poor populations. Studies across a number of developing nations have shown that those who receive exclusive breastfeeding during their first 6 months of life are better protected against infection with diarrheal diseases. Exclusive breastfeeding is currently recommended during, at least, the first six months of an infant's life by the WHO.

Others

Probiotics decrease the risk of diarrhea in those taking antibiotics.

Management

In many cases of diarrhea, replacing lost fluid and salts is the only treatment needed. This is usually by mouth – oral rehydration therapy – or, in severe cases, intravenously. Diet restrictions such as the BRAT diet are no longer recommended. Research does not support the limiting of milk to children as doing so has no effect on duration of diarrhea. To the contrary, WHO recommends that children with diarrhea continue to eat as sufficient nutrients are usually still absorbed to support continued growth and weight gain, and that continuing to eat also speeds up recovery of normal intestinal functioning. CDC recommends that children and adults with cholera also continue to eat.

Medications such as loperamide (Imodium) and bismuth subsalicylate may be beneficial; however they may be contraindicated in certain situations.

Fluids

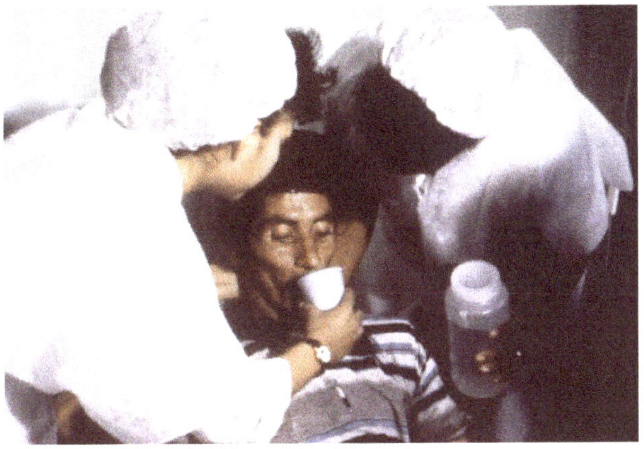

A person consuming oral rehydration solution.

Oral rehydration solution (ORS) (a slightly sweetened and salty water) can be used to prevent dehydration. Standard home solutions such as salted rice water, salted yogurt drinks, vegetable and chicken soups with salt can be given. Home solutions such as water in which cereal has been cooked, unsalted soup, green coconut water, weak tea (unsweetened), and unsweetened fresh fruit juices can have from half a teaspoon to full teaspoon of salt (from one-and-a-half to three grams) added per liter. Clean plain water can also be one of several fluids given. There are commercial solutions such as Pedialyte, and relief agencies such as UNICEF widely distribute packets of salts and sugar. A WHO publication for physicians recommends a homemade ORS consisting of one liter water with one teaspoon salt (3 grams) and two tablespoons sugar (18 grams) added (approximately the "taste of tears"). Rehydration Project recommends adding the same amount of sugar but only one-half a teaspoon of salt, stating that this more dilute approach is less risky with very little loss of effectiveness. Both agree that drinks with too much sugar or salt can make dehydration worse.

Appropriate amounts of supplemental zinc and potassium should be added if available. But the availability of these should not delay rehydration. As WHO points out, the most important thing is to begin preventing dehydration as early as possible. In another example of prompt ORS hopefully preventing dehydration, CDC recommends for the treatment of cholera continuing to give Oral Rehydration Solution during travel to medical treatment.

Vomiting often occurs during the first hour or two of treatment with ORS, especially if a child drinks the solution too quickly, but this seldom prevents successful rehydration since most of the fluid is still absorbed. WHO recommends that if a child vomits, to wait five or ten minutes and then start to give the solution again more slowly.

Drinks especially high in simple sugars, such as soft drinks and fruit juices, are not recommended in children under 5 years of age as they may *increase* dehydration. A too rich solution in the gut draws water from the rest of the body, just as if the person were to drink sea water. Plain water may be used if more specific and effective ORT preparations are unavailable or are not palatable. Additionally, a mix of both plain water and drinks perhaps too rich in sugar and salt can alternatively be given to the same person, with the goal of providing a medium amount of sodium overall. A nasogastric tube can be used in young children to administer fluids if warranted.

Eating

WHO recommends a child with diarrhea continue to be fed. Continued feeding speeds the recovery of normal intestinal function. In contrast, children whose food is restricted have diarrhea of longer duration and recover intestinal function more slowly. A child should also continue to be breastfed. The WHO states "Food should *never* be withheld and the child's usual foods should *not* be diluted. Breastfeeding should *always* be continued." And in the specific example of cholera, CDC also makes the same recommendation. In young children who are not breast-fed and live in the developed world, a lactose-free diet may be useful to speed recovery.

Medications

While antibiotics are beneficial in certain types of acute diarrhea, they are usually not used except in specific situations. There are concerns that antibiotics may increase the risk of hemolytic uremic syndrome in people infected with Escherichia coli O157:H7. In resource-poor countries, treatment with antibiotics may be beneficial. However, some bacteria are developing antibiotic resistance, particularly *Shigella*. Antibiotics can also cause diarrhea, and antibiotic-associated diarrhea is the most common adverse effect of treatment with general antibiotics.

While bismuth compounds (Pepto-Bismol) decreased the number of bowel movements in those with travelers' diarrhea, they do not decrease the length of illness. Anti-motility agents like loperamide are also effective at reducing the number of stools but not the duration of disease. These agents should only be used if bloody diarrhea is not present.

Bile acid sequestrants such as cholestyramine can be effective in chronic diarrhea due to bile acid malabsorption. Therapeutic trials of these drugs are indicated in chronic diarrhea if bile acid malabsorption cannot be diagnosed with a specific test, such as SeHCAT retention.

Alternative Therapies

Zinc supplementation benefits children with diarrhea in developing countries, but only in infants over six months old. This supports the World Health Organization guidelines for zinc, but not in the very young.

Probiotics reduce the duration of symptoms by one day and reduced the chances of symptoms lasting longer than four days by 60%. The probiotic lactobacillus can help prevent antibiotic-associated diarrhea in adults but possibly not children. For those with lactose intolerance, taking digestive enzymes containing lactase when consuming dairy products often improves symptoms.

Epidemiology

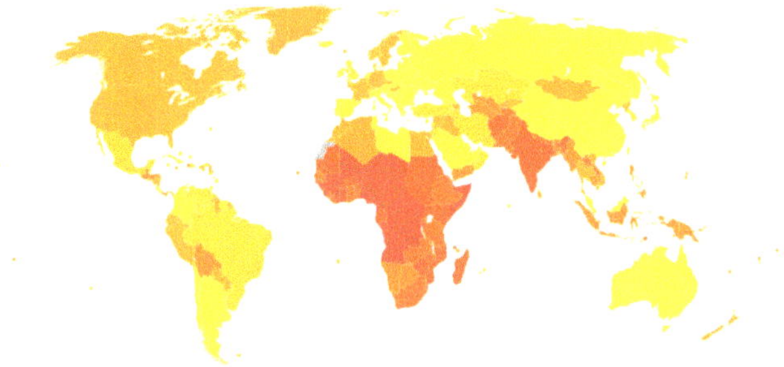

Deaths due to diarrhoeal diseases per million persons in 2012

- 0-2
- 3-10
- 11-18
- 19-30
- 31-46
- 47-80
- 81-221
- 222-450
- 451-606
- 607-1799

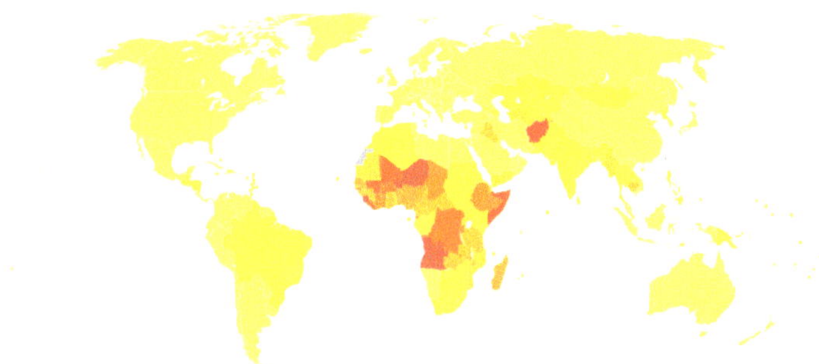

Disability-adjusted life year for diarrhea per 100,000 inhabitants in 2004.

- no data
- < 500
- 500–1000
- 1000–1500
- 1500–2000

- 2000–2500
- 2500–3000
- 3000–3500
- 3500–4000
- 4000–4500
- 4500–5000
- 5000–6000
- > 6000

Worldwide in 2004, approximately 2.5 billion cases of diarrhea occurred, which resulted in 1.5 million deaths among children under the age of five. Greater than half of these were in Africa and South Asia. This is down from a death rate of 4.5 million in 1980 for gastroenteritis. Diarrhea remains the second leading cause of infant mortality (16%) after pneumonia (17%) in this age group.

The majority of such cases occur in the developing world, with over half of the recorded cases of childhood diarrhea occurring in Africa and Asia, with 696 million and 1.2 billion cases, respectively, compared to only 480 million in the rest of the world.

Infectious diarrhea resulted in about 0.7 million deaths in children under five years old in 2011 and 250 million lost school days. In the Americas, diarrheal disease accounts for a total of 10% of deaths among children aged 1–59 months while in South East Asia, it accounts for 31.3% of deaths. It is estimated that around 21% of child mortalities in developing countries are due to diarrheal disease.

Etymology

The word diarrhea is from "through" and *rheo* "flow". Diarrhea is the spelling in American English while diarrhoea is the spelling in Commonwealth English.

Nausea

Nausea is a sensation of unease and discomfort in the upper stomach with an involuntary urge to vomit. It may precede vomiting, but a person can have nausea without vomiting. When prolonged, it is a debilitating symptom.

Nausea is a non-specific symptom, which means that it has many possible causes. Some common causes of nausea are motion sickness, dizziness, migraine, fainting, gastroenteritis (stomach infection) or food poisoning. Nausea is a side effect of many medications including chemotherapy, nauseants or morning sickness in early pregnancy. Nausea may also be caused by anxiety, disgust and depression.

Medications taken to prevent and treat nausea are called antiemetics. The most commonly prescribed antiemetics in the US are promethazine, metoclopramide and ondansetron. The word nausea is from Latin *nausea*, “motion sickness”, “feeling sick or queasy”.

Causes

There are many causes of nausea. One organization listed 700 in 2009. Gastrointestinal infections (37%) and food poisoning are the two most common causes. Side effects from medications (3%) and pregnancy are also relatively frequent. In 10% of people the cause remains unknown.

Food Poisoning

Food poisoning usually causes an abrupt onset of nausea and vomiting one to six hours after ingestion of contaminated food and lasts for one to two days. It is due to toxins produced by bacteria in food.

Medications

Many medications can potentially cause nausea. Some of the most frequently associated include cancer and systemic chemotherapy regimens and general anaesthetic agents.

Pregnancy

Nausea or “morning sickness” is common during early pregnancy but may occasionally continue into the second and third trimesters. In the first trimester nearly 80% of women have some degree of nausea. Pregnancy should therefore be considered as a possible cause of nausea in any women of child bearing age. While usually it is mild and self-limiting, severe cases known as hyperemesis gravidarum may require treatment.

Disequilibrium

A number of conditions involving balance such as motion sickness and vertigo can lead to nausea and vomiting.

Stress and Depression

Nausea may be caused by stress and depression.

Potentially Serious

While most causes of nausea are not serious, some serious causes do occur. These include: Intracranial Pressure secondary to head trauma or hemorrhagic stroke, diabetic ketoacidosis, brain tumor, surgical problems, heart attack, pancreatitis, small bowel obstruction, meningitis, appendicitis, cholecystitis, Addisonian crisis, Choledocholithiasis (from gallstones), hepatitis, as a sign of carbon monoxide poison and many others.

Comprehensive List

Inside the Abdomen

Obstructing Disorders

- Pyloric obstruction
- Small bowel obstruction
- Colonic obstruction
- Superior mesenteric artery syndrome

Enteric Infections

- Viral infection
- Bacterial infection

Inflammatory Diseases

- Cholecystitis
- Pancreatitis
- Appendicitis
- Hepatitis

Sensorimotor Dysfunction

- Gastroparesis
- Intestinal pseudo-obstruction
- Gastroesophageal reflux disease
- Chronic idiopathic nausea
- Functional vomiting
- Cyclic vomiting syndrome
- Rumination syndrome

Other

- Biliary colic
- Abdominal irradiation

Outside the Abdomen

Cardiopulmonary

- Cardiomyopathy

- Myocardial infarction (heart attack)

Inner-ear Diseases

- Motion sickness
- Labyrinthitis
- Malignancy

Intracerebral Disorders

- Malignancy
- Hemorrhage
- Abscess
- Hydrocephalus

Psychiatric Illnesses

- Anorexia and bulimia nervosa
- Depression

Other

- Post-operative vomiting

Medications and Metabolic Disorders

Drugs

- Chemotherapy
- Antibiotics
- Antiarrhythmics
- Digoxin
- Oral hypoglycemic medications
- Oral contraceptives

Endocrine/Metabolic Disease

- Pregnancy
- Uremia
- Ketoacidosis
- Thyroid and parathyroid disease
- Adrenal insufficiency

Toxins

- Liver failure
- Alcohol

Diagnostic Approach

Patient History

Taking a thorough patient history may reveal important clues to the cause of nausea and vomiting. If the patient's symptoms have an acute onset, then drugs, toxins, and infections are likely. In contrast, a long-standing history of nausea will point towards a chronic illness as the culprit. The timing of nausea and vomiting after eating food is an important factor to pay attention to. Symptoms that occur within an hour of eating may indicate an obstruction proximal to the small intestine, such as gastroparesis or pyloric stenosis. An obstruction further down in the intestine or colon will cause delayed vomiting. An infectious cause of nausea and vomiting such as gastroenteritis may present several hours to days after the food was ingested. The contents of the emesis is a valuable clue towards determining the cause. Bits of fecal matter in the emesis indicate obstruction in the distal intestine or the colon. Emesis that is of a bilious nature (greenish in color) localizes the obstruction to a point past the stomach. Emesis of undigested food points to an obstruction prior to the gastric outlet, such as achalasia or Zenker's diverticulum. If patient experiences reduced abdominal pain after vomiting, then obstruction is a likely etiology. However, vomiting does not relieve the pain brought on by pancreatitis or cholecystitis.

Physical Exam

It is important to watch out for signs of dehydration, such as orthostatic hypotension and skin turgor. Auscultation of the abdomen can produce several clues to the cause of nausea and vomiting. A high-pitched tinkling sound indicates possible bowel obstruction, while a splashing "succussion" sound is more indicative of gastric outlet obstruction. Eliciting pain on the abdominal exam when pressing on the patient may indicate an inflammatory process. Signs such as papilledema, visual field losses, or focal neurological deficits are red flag signs for elevated intracranial pressure.

Diagnostic Testing

When a history and physical exam are not enough to determine the cause of nausea and vomiting, certain diagnostic tests may prove useful. A chemistry panel would be useful for electrolyte and metabolic abnormalities. Liver function tests and lipase would identify pancreaticobiliary diseases. Abdominal X-rays showilng air-fluid levels indicate bowel obstruction, while an X-ray showing air-filled bowel loops are more indicative of ileus. More advanced imaging and procedures may be necessary, such as a CT scan, upper endoscopy, colonoscopy, barium enema, or MRI. Abnormal GI motility can be assessed using specific tests like gastric scintigraphy, wireless motility capsules, and small-intestinal manometry.

Pathophysiology

Research on nausea and vomiting has relied on using animal models to mimic the anatomy and neuropharmacologic features of the human body. The physiologic mechanism of nausea is a complex process that has yet to be fully elucidated. There are four general pathways that are activated by specific triggers in the human body that go on to create the sensation of nausea and vomiting.

- Central nervous system (CNS): Stimuli can affect areas of the CNS including the cerebral cortex and the limbic system. These areas are activated by elevated intracranial pressure, irritation of the meninges (i.e. blood or infection), and extreme emotional triggers such as anxiety.
- Chemoreceptor trigger zone (CTZ): The CTZ is located in the area postrema in the floor of the fourth ventricle within the brain. This area is outside the blood brain barrier, and is therefore readily exposed to substances circulating through the blood and cerebral spinal fluid. Common triggers of the CTZ include metabolic abnormalities, toxins, and medications. Activation of the CTZ is mediated by dopamine (D2) receptors, serotonin (5HT3) receptors, and neurokinin receptors (NK1).
- Vestibular system: This system is activated by disturbances to the vestibular apparatus in the inner ear. These include movements that cause motion sickness and dizziness. This pathway is triggered via histamine (H1) receptors and acetylcholine (ACh) receptors.
- Peripheral Pathways: These pathways are triggered via chemoreceptors and mechanoreceptors in the gastrointestinal tract, as well as other organs such as the heart and kidneys. Common activators of these pathways include toxins present in the gastrointestinal lumen and distension of the gastrointestinal lumen from blockage or dysmotility of the bowels. Signals from these pathways travel via multiple neural tracts including the vagus, glossopharyngeal, splanchnic, and sympathetic nerves.

Signals from any of these pathways then travel to the brainstem, activating several structures including the nucleus of the solitary tract, the dorsal motor nucleus of the vagus, and central pattern generator. These structures go on to signal various downstream effects of nausea and vomiting. The body's motor muscle responses involve halting the muscles of the gastrointestinal tract, and in fact causing reversed propulsion of gastric contents towards the mouth while increasing abdominal muscle contraction. Autonomic effects involve increased salivation and the sensation of feeling faint that often occurs with nausea and vomiting.

Treatment

If dehydration is present due to loss of fluids from severe vomiting, rehydration with oral electrolyte solutions is preferred. If this is not effective or possible, intravenous rehydration may be required. Medical care is recommended if: a person cannot keep any liquids down, has symptoms more than 2 days, is weak, has a fever, has stomach pain, vomits more two times in a day or does not urinate for more than 8 hours.

Medications

Many pharmacologic medications are available for the treatment of nausea. There is no medication that is clearly superior to other medications for all cases of nausea. The choice of antiemetic medication may be based on the situation during which the person experiences nausea. For people with motion sickness and vertigo, antihistamines and anticholinergics such as meclizine and scopalamine are particularly effective. Nausea and vomiting associated with migraine headaches respond best to dopamine antagonists such as metoclopramide, prochlorperazine, and chlorpromazine. In cases of gastroenteritis, serotonin antagonist such as ondansetron were found to suppress nausea and vomiting, as well as reduce the need for IV fluid resuscitation. The combination of pyridoxine and doxylamine is the first line treatment for pregnancy-related nausea and vomiting. Dimenhydrinate is an inexpensive and effective over the counter medication for preventing postoperative nausea and vomiting. Other factors to consider when choosing an antiemetic medication include the person's preference, side-effect profile, and cost.

Alternative Medicine

In certain people, cannabinoids may be effective in reducing chemotherapy associated nausea and vomiting. Several studies have demonstrated the therapeutic effects of cannabinoids for nausea and vomiting in the advanced stages of illnesses such as cancer and AIDS.

In hospital settings topical anti-nausea gels are not indicated because of lack of research backing their efficacy. Topical gels containing lorazepam, diphenhydramine, and haloperidol are sometimes used for nausea but are not equivalent to more established therapies.

Ginger has also been shown to be potentially effective in treating several types of nausea.

Tentative evidence supports acupuncture at point PC6.

Prognosis

The outlook depends on the cause. Most people recover within few hours or a day. While short-term nausea and vomiting are generally harmless, they may sometimes indicate a more serious condition. When associated with prolonged vomiting, it may lead to dehydration and/or dangerous electrolyte imbalances. Repeated intentional vomiting, characteristic of bulimia, can cause stomach acid to wear away at the enamel in teeth.

Epidemiology

Nausea and or vomiting is the main complaint in 1.6% of visits to family physicians in Australia. However, only 25% of people with nausea visit their family physician. It is most common in those 15–24 years old and less common in other ages.

Vomiting

Vomiting, also known as emesis and throwing up, among other terms, is the involuntary, forceful expulsion of the contents of one's stomach through the mouth and sometimes the nose.

Vomiting can be caused by a wide variety of conditions; it may present as a specific response to ailments like gastritis or poisoning, or as a non-specific sequela of disorders ranging from brain tumors and elevated intracranial pressure to overexposure to ionizing radiation. The feeling that one is about to vomit is called nausea, which often precedes, but does not always lead to, vomiting. Antiemetics are sometimes necessary to suppress nausea and vomiting. In severe cases, where dehydration develops, intravenous fluid may be required.

Vomiting is different from regurgitation, although the two terms are often used interchangeably. Regurgitation is the return of undigested food back up the esophagus to the mouth, without the force and displeasure associated with vomiting. The causes of vomiting and regurgitation are generally different.

Complications

Aspiration of Vomit

Vomiting can be dangerous if the gastric content enters the respiratory tract. Under normal circumstances the gag reflex and coughing prevent this from occurring; however, these protective reflexes are compromised in persons under the influences of certain substances such as alcohol or anesthesia. The individual may choke and asphyxiate or suffer an aspiration pneumonia.

Dehydration and Electrolyte Imbalance

Prolonged and excessive vomiting depletes the body of water (dehydration), and may alter the electrolyte status. Gastric vomiting leads to the loss of acid (protons) and chloride directly. Combined with the resulting alkaline tide, this leads to hypochloremic metabolic alkalosis (low chloride levels together with high HCO−3 and CO2 and increased blood pH) and often hypokalemia (potassium depletion). The hypokalemia is an indirect result of the kidney compensating for the loss of acid. With the loss of intake of food the individual may eventually become cachectic. A less frequent occurrence results from a vomiting of intestinal contents, including bile acids and HCO−3, which can cause metabolic acidosis.

Mallory–weiss Tear

Repeated or profuse vomiting may cause erosions to the esophagus or small tears in the esophageal mucosa (Mallory–Weiss tear). This may become apparent if fresh red blood is mixed with vomit after several episodes.

Dentistry

Recurrent vomiting, such as observed in bulimia nervosa, may lead to destruction of the tooth enamel due to the acidity of the vomit. Digestive enzymes can also have a negative effect on oral health, by degrading the tissue of the gums.

Pathophysiology

Receptors on the floor of the fourth ventricle of the brain represent a chemoreceptor trigger zone, known as the area postrema, stimulation of which can lead to vomiting. The area postrema is a circumventricular organ and as such lies outside the blood–brain barrier; it can therefore be stimulated by blood-borne drugs that can stimulate vomiting or inhibit it.

14th-century illustration of vomiting from the Casanatense *Tacuinum Sanitatis*

There are various sources of input to the vomiting center:

- The chemoreceptor trigger zone at the base of the fourth ventricle has numerous dopamine D_2 receptors, serotonin 5-HT_3 receptors, opioid receptors, acetylcholine receptors, and receptors for substance P. Stimulation of different receptors are involved in different pathways leading to emesis, in the final common pathway substance P appears involved.
- The vestibular system, which sends information to the brain via cranial nerve VIII (vestibulocochlear nerve), plays a major role in motion sickness, and is rich in muscarinic receptors and histamine H_1 receptors.
- The cranial nerve X (vagus nerve) is activated when the pharynx is irritated, leading to a gag reflex.
- The vagal and enteric nervous system inputs transmit information regarding the state of the gastrointestinal system. Irritation of the GI mucosa by chemotherapy, radiation, distention, or acute infectious gastroenteritis activates the 5-HT_3 receptors of these inputs.
- The CNS mediates vomiting that arises from psychiatric disorders and stress from higher brain centers.

The vomiting act encompasses three types of outputs initiated by the chemoreceptor trigger zone: Motor, parasympathetic nervous system (PNS), and sympathetic nervous system (SNS). They are as follows:

- Increased salivation to protect tooth enamel from stomach acids. (Excessive vomiting leads to dental erosion). This is part of the PNS output.
- The body takes a deep breath to avoid aspirating vomit.
- Retroperistalsis starts from the middle of the small intestine and sweeps up digestive tract contents into the stomach, through the relaxed pyloric sphincter.

- Intrathoracic pressure lowers (by inspiration against a closed glottis), coupled with an increase in abdominal pressure as the abdominal muscles contract, propels stomach contents into the esophagus as the lower esophageal sphincter relaxes. The stomach itself does not contract in the process of vomiting except for at the angular notch, nor is there any retroperistalsis in the esophagus.
- Vomiting is ordinarily preceded by retching.
- Vomiting also initiates an SNS response causing both sweating and increased heart rate.

The neurotransmitters that regulate vomiting are poorly understood, but inhibitors of dopamine, histamine, and serotonin are all used to suppress vomiting, suggesting that these play a role in the initiation or maintenance of a vomiting cycle. Vasopressin and neurokinin may also participate.

Phases

The vomiting act has two phases. In the retching phase, the abdominal muscles undergo a few rounds of coordinated contractions together with the diaphragm and the muscles used in respiratory inspiration. For this reason, an individual may confuse this phase with an episode of violent hiccups. In this retching phase nothing has yet been expelled. In the next phase, also termed the expulsive phase, intense pressure is formed in the stomach brought about by enormous shifts in both the diaphragm and the abdomen. These shifts are, in essence, vigorous contractions of these muscles that last for extended periods of time - much longer than a normal period of muscular contraction. The pressure is then suddenly released when the upper esophageal sphincter relaxes resulting in the expulsion of gastric contents. Individuals who do not regularly exercise their abdominal muscles may experience pain in those muscles for a few days. The relief of pressure and the release of endorphins into the bloodstream after the expulsion causes the vomiter to feel better.

Contents

Gastric secretions and likewise vomit are highly acidic. Recent food intake appears in the gastric vomit. Irrespective of the content, vomit tends to be malodorous.

The content of the vomitus (vomit) may be of medical interest. Fresh blood in the vomit is termed hematemesis ("blood vomiting"). Altered blood bears resemblance to coffee grounds (as the iron in the blood is oxidized) and, when this matter is identified, the term coffee-ground vomiting is used. Bile can enter the vomit during subsequent heaves due to duodenal contraction if the vomiting is severe. Fecal vomiting is often a consequence of intestinal obstruction or a gastrocolic fistula and is treated as a warning sign of this potentially serious problem (*signum mali ominis*).

If the vomiting reflex continues for an extended period with no appreciable vomitus, the condition is known as non-productive emesis or "dry heaves", which can be painful and debilitating.

Color of Vomit

- Bright red in the vomit suggests bleeding from the esophagus

- Dark red vomit with liver-like clots suggests profuse bleeding in the stomach, such as from a perforated ulcer
- Coffee-ground-like vomit suggests less severe bleeding in the stomach, because the gastric acid has had time to change the composition of the blood
- Yellow vomit suggests bile, indicating that the pyloric valve is open and bile is flowing into the stomach from the duodenum (this is more common in older people)

Differential Diagnosis

Vomiting may be due to a large number of causes, and protracted vomiting has a long differential diagnosis.

Digestive Tract

Causes in the digestive tract

- Gastritis (inflammation of the gastric wall)
- Gastroenteritis
- Gastroesophageal reflux disease
- Pyloric stenosis (in babies, this typically causes a very forceful "projectile vomiting" and is an indication for urgent surgery)
- Bowel obstruction
- Overeating
- Acute abdomen and/or peritonitis
- Ileus
- Food allergies (often in conjunction with hives or swelling)
- Cholecystitis, pancreatitis, appendicitis, hepatitis
- Food poisoning
- In children, it can be caused by an allergic reaction to cow's milk proteins (Milk allergy or lactose intolerance)

Sensory System and Brain

Causes in the sensory system:

- Movement: motion sickness (which is caused by overstimulation of the labyrinthine canals of the ear)
- Ménière's disease

An intoxicated woman vomiting

Causes in the brain:

- Concussion
- Cerebral hemorrhage
- Migraine
- Brain tumors, which can cause the chemoreceptors to malfunction
- Benign intracranial hypertension and hydrocephalus

Metabolic disturbances (these may irritate both the stomach and the parts of the brain that coordinate vomiting):

- Hypercalcemia (high calcium levels)
- Uremia (urea accumulation, usually due to renal failure)
- Adrenal insufficiency
- Hypoglycemia
- Hyperglycemia

Pregnancy:

- Hyperemesis, morning sickness

Drug reaction (vomiting may occur as an acute somatic response to):

- alcohol (being sick while being drunk or being sick the next morning, suffering from the after-effects, i.e., the hangover)
- opioids
- selective serotonin reuptake inhibitors
- many chemotherapy drugs
- some entheogens (such as peyote or ayahuasca)

Illness (sometimes colloquially known as "stomach flu"—a broad name that refers to gastric inflammation caused by a range of viruses and bacteria):

- Norovirus (formerly Norwalk virus or Norwalk agent)
- Swine influenza

Emetics

An *emetic*, such as syrup of ipecac, is a substance that induces vomiting when administered orally or by injection. An emetic is used medically when a substance has been ingested and must be expelled from the body immediately (for this reason, many toxic and easily digestible products such as rat poison contain an emetic). Inducing vomiting can remove the substance before it is absorbed into the body. Ipecac abuse can cause detrimental health effects.

Salt water and mustard water have been used since ancient times as emetics. Care must be taken with salt, as excessive intake can potentially be harmful.

Copper sulfate was also used in the past as an emetic. It is now considered too toxic for this use.

Hydrogen peroxide is used as an emetic in veterinary practice.

Social Cues

A drunk man vomiting, while a young slave is holding his forehead. Brygos Painter, 500–470 BC

It is quite common that, when one person vomits, others nearby become nauseated, particularly when smelling the vomit of others, often to the point of vomiting themselves. It is believed that this is an evolved trait among primates. Many primates in the wild tend to browse for food in small groups. Should one member of the party react adversely to some ingested food, it may be advantageous (in a survival sense) for other members of the party to also vomit. This tendency in human populations has been observed at drinking parties, where excessive consumption of alcoholic beverages may cause a number of party members to vomit nearly simultaneously, this being triggered by the initial vomiting of a single member of the party. This phenomenon has been touched on in popular culture: notorious instances appear in the films *Monty Python's The Meaning of Life* (1983) and *Stand By Me* (1986).

Intense vomiting in ayahuasca ceremonies is a common phenomenon. However, people who experience "la purga" after drinking ayahuasca, in general, regard the practice as both a physical and spiritual cleanse and often come to welcome it. It has been suggested that the consistent emetic effects of ayahuasca—in addition to its many other therapeutic properties—was of medicinal benefit to indigenous peoples of the Amazon, in helping to clear parasites from the gastrointestinal system.

There have also been documented cases of a single ill and vomiting individual inadvertently causing others to vomit, when they are especially fearful of also becoming ill, through a form of mass hysteria.

Special bags are often supplied on boats for sick passengers to vomit into.

Most people try to contain their vomit by vomiting into a sink, toilet, or trash can, as vomit is difficult and unpleasant to clean. On airplanes and boats, special bags are supplied for sick passengers to vomit into. A special disposable bag (leakproof, puncture-resistant, odorless) containing absorbent material that solidifies the vomit quickly is also available, making it convenient and safe to store until there is an opportunity to dispose of it conveniently.

People who vomit chronically (e.g., as part of an eating disorder such as bulimia nervosa) may devise various ways to hide this disorder.

An online study of people's responses to "horrible sounds" found vomiting "the most disgusting".

Professor Trevor Cox of the University of Salford's Acoustic Research Centre said that "We are pre-programmed to be repulsed by horrible things such as vomiting, as it is fundamental to staying alive to avoid nasty stuff." It is thought that disgust is triggered by the sound of vomiting to protect those nearby from possibly diseased food.

Miscellanea

- Self-induced
  - Eating disorders (anorexia nervosa or bulimia nervosa)
  - To eliminate an ingested poison (some poisons should not be vomited as they may be more toxic when inhaled or aspirated; it is better to ask for help before inducing vomiting)
  - Some people who engage in binge drinking induce vomiting to make room in their stomachs for more alcohol consumption.
  - People suffering from nausea may induce vomiting in hopes of feeling better.
- After surgery (postoperative nausea and vomiting)
- Disagreeable sights or disgust, smells or thoughts (such as decayed matter, others' vomit, thinking of vomiting), etc.
- Extreme pain, such as intense headache or myocardial infarction (heart attack)
- Violent emotions
- Cyclic vomiting syndrome (a poorly understood condition with attacks of vomiting)
- High doses of ionizing radiation sometimes trigger a vomit reflex.
- Violent fits of coughing, hiccups, or asthma
- Anxiety
- Depression
- Overexertion (doing too much strenuous exercise can lead to vomiting shortly afterwards).
- Rumination syndrome, an underdiagnosed and poorly understood disorder that causes sufferers to regurgitate food shortly after ingestion.

Other Types

- Projectile vomiting refers to vomiting that ejects the gastric contents with great force. It is a classic symptom of infantile hypertrophic pyloric stenosis, in which it typically follows feeding and can be so forceful that some material exits through the nose.

Treatment

An antiemetic is a drug that is effective against vomiting and nausea. Antiemetics are typically used to treat motion sickness and the side effects of medications such as opioids and chemotherapy.

Antiemetics act by inhibiting the receptor sites associated with emesis. Hence, anticholinergics, antihistamines, dopamine antagonists, serotonin antagonists, and cannabinoids are used as antiemetics.

Evidence to support the use of antiemetics for nausea and vomiting among adults in the emergency department is poor. It is unclear if any medication is better than another or better than no active treatment.

Epidemiology

Nausea and/or vomiting are the main complaints in 1.6% of visits to family physicians in Australia.

References

- Smith, David F.; Diack, H. Lesley and Pennington, T. Hugh (2005) Food Poisoning, Policy and Politics : Corned Beef and Typhoid in Britain in the 1960s. Boydell Press. ISBN 1-84383-138-4
- Shors, Teri (2013). The microbial challenge : a public health perspective (3rd ed.). Burlington, Mass.: Jones & Bartlett Learning. p. 457. ISBN 9781449673338.
- A. Helms, Richard (2006). Textbook of therapeutics : drug and disease management (8. ed.). Philadelphia, Pa. [u.a.]: Lippincott Williams & Wilkins. p. 2003. ISBN 9780781757348.
- Webber, Roger (2009). Communicable disease epidemiology and control : a global perspective (3rd ed.). Wallingford, Oxfordshire: Cabi. p. 79. ISBN 978-1-84593-504-7.
- Dolin, [edited by] Gerald L. Mandell, John E. Bennett, Raphael (2010). "93". Mandell, Douglas, and Bennett's principles and practice of infectious diseases (7th ed.). Philadelphia, PA: Churchill Livingstone/Elsevier. ISBN 0-443-06839-9.
- Tintinalli, Judith E. (2010). Emergency Medicine: A Comprehensive Study Guide (Emergency Medicine (Tintinalli)). New York: McGraw-Hill Companies. pp. 830–839. ISBN 0-07-148480-9.
- Mandell, Gerald L.; Bennett, John E.; Dolin, Raphael (2004). Mandell's Principles and Practices of Infection Diseases (6th ed.). Churchill Livingstone. ISBN 0-443-06643-4.
- Feldman, Mark; Friedman, Lawrence S.; Sleisenger, Marvin H. (2002). Sleisenger & Fordtran's Gastrointestinal and Liver Disease (7th ed.). Saunders. ISBN 0-7216-8973-6.
- Charles E. Rosenberg (2009). The Cholera Years the United States in 1832, 1849, and 1866. Chicago: University of Chicago Press. p. 74. ISBN 9780226726762.
- Rousseaux, Wanda Haschek, Matthew Wallig, Colin (2009). Fundamentals of toxicologic pathology (2nd ed.). London: Academic. p. 182. ISBN 9780123704696.
- MacLachlan, edited by N. James; Dubovi, Edward J. (2009). Fenner's veterinary virology (4th ed.). Amsterdam: Elsevier Academic Press. p. 399. ISBN 9780123751584.
- Zimmerman, Jeffrey; Karriker, Locke; Ramirez, Alejandro; Kent Schwartz; Gregory Stevenson (15 May 2012). Diseases of Swine (10th ed.). Chichester, West Sussex: John Wiley & Sons. p. 504. ISBN 978-0-8138-2267-9.
- Basem Abdelmalak; John Doyle, eds. (2013). Anesthesia for otolaryngologic surgery. Cambridge University Press. pp. 282–287. ISBN 1107018676.
- Prober, edited by Sarah Long, Larry Pickering, Charles G. (2012). Principles and practice of pediatric infectious diseases (4th ed.). Edinburgh: Elsevier Saunders. p. 96. ISBN 9781455739851.
- Kasper DL, Braunwald E, Fauci AS, Hauser SL, Longo DL, Jameson JL. Harrison's Principles of Internal Medicine. New York: McGraw-Hill, 2005. ISBN 0-07-139140-1.
- Tintinalli, Judith E. (2010). Emergency Medicine: A Comprehensive Study Guide (Emergency Medicine (Tin-

tinalli)). New York: McGraw-Hill Companies. p. 830. ISBN 0-07-148480-9.

- Holland, James F.; Kufe, Donald W.; Weichselbaum, Ralph R.; Pollock, Raphael E.; Frei III, Emil; Gansler, Ted S.; Bast Jr., Robert C. (2003). Cancer medicine (6. [ed.]. ed.). Hamilton, Ontario [u.a.]: Decker. ISBN 9781550092134.
- Sleisenger, edited by Mark Feldman, Lawrence S. Friedman, Lawrence J. Brandt; consulting editor, Marvin H. (2009). Sleisenger & Fordtran's gastrointestinal and liver disease pathophysiology, diagnosis, management (PDF) (9th ed.). St. Louis, Mo.: MD Consult. p. 783. ISBN 1-4160-6189-4.

6

Food Quality and Related Aspects

The inspection of food to make sure its fresh and to confirm its market value is food grading whereas the process of adding micronutrients to food items is food fortification. This chapter elucidates on topics such as, food grading, food fortification, optical sorting and food intolerance.

Food Grading

Food grading involves the inspection, assessment and sorting of various foods regarding quality, freshness, legal conformity and market value. Food grading often occurs by hand, in which foods are assessed and sorted. Machinery is also used to grade foods, and may involve sorting products by size, shape and quality. For example, machinery can be used to remove spoiled food from fresh product.

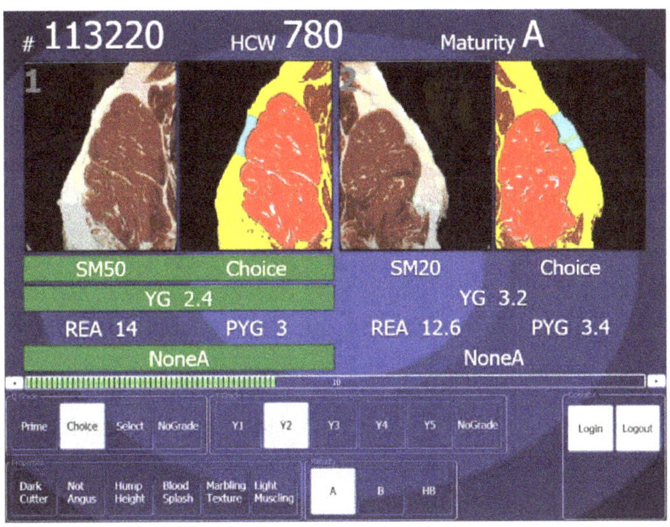

A screenshot from the electronic grading system showing USDA Choice, Yield Grade 2 beef. The left is the natural color view of the cut; the right is the instrument enhanced view that details the amount of marbling, size, and fat thickness.

By Food Type

Beef

Beef grading in the United States is performed by the United States Department of Agriculture's (USDA) Agricultural and Marketing Service. There are eight beef quality grades, with U.S. Prime being the highest grade and U.S. Canner being the lowest grade. Beef grading is a complex and detailed process, requiring graders to think and calculate quickly with great accuracy. Using technology to complement and supplement the onsite human graders generates an efficient and more precise process.

Inspected beef carcasses tagged by the USDA

Beer

In beer grading, the letter "X" is used on some beers, and was traditionally a mark of beer strength, with the more Xs the greater the strength. Some sources suggest that the origin of the mark was in the breweries of medieval monasteries Another plausible explanation is contained in a treatise entitled "The Art of Brewing" published in London in 1829. It says; *"The duties on ale and beer, which were first imposed in 1643... at a certain period, in distinguishing between small beer and strong, all ale or beer, sold at or above ten shillings per barrel, was reckoned to be* strong *and was, therefore, subjected to a higher duty. The cask which contained this strong beer was then first marked with an X signifying ten; and hence the present quack-like denominations of XX (double X) and XXX (treble X) on the casks and accounts of the strong-ale brewers".*

In mid-19th century England, the use of "X" and other letters had evolved into a standardised grading system for the strength of beer. Today, it is used as a trade mark by a number of brewers in the United Kingdom, the Commonwealth and the United States.

European Bitterness Units scale, often abbreviated as EBU, is a scale for measuring the perceived bitterness of beer, with lower values being generally "less bitter" and higher values "more bitter". The scale and method are defined by the European Brewery Convention, and the numerical value should be the same as of the International Bitterness Units scale (IBU), defined in co-operation with the American Society of Brewing Chemists. However, the exact process of determining EBU and IBU values differs slightly, which may in theory result with slightly smaller values for EBU than IBU.

The International Bittering Units scale, or simply IBU scale, provides a measure of the bitterness of beer, which is provided by the hops used during brewing. Bittering units are measured through the use of a spectrophotometer and solvent extraction.

Butter

Coconut Milk

Several grades of coconut milk exist: from thick at 20-22% fat to thin at 5-7% fat level.

Coffee

Dark-roasted coffee beans

Coffee growers, traders, and roasters grade beans based on size, color, and a variety of other characteristics. Coffees of exceptional quality are traded as "specialty coffees" and fetch a higher price in the international market.

After the roast, Coffee grading involves assessment of roasted coffee seed colorization and then labeling as light, medium light, medium, medium dark, dark, or very dark. A more accurate method of discerning the degree of roast involves measuring the reflected light from roasted seeds illuminated with a light source in the near infrared spectrum. This elaborate light meter uses a process known as spectroscopy to return a number that consistently indicates the roasted coffee's relative degree of roast or flavor development.

Eggs

In the United States, egg grading is performed by the USDA, and is based upon the interior quality of the egg and the appearance and condition of the egg shell. Eggs of any quality grade may differ in weight (size). Egg grading is performed by candling, which involves observing the interior of eggs by placing them in front of a bright light.

Guar Gum

Guar gum grading involves analysis for coloration, viscosity, moisture, granulation, protein content and insolubles ash.

Honey

Honey grading in the United States is performed voluntarily based upon USDA standards (USDA does offer inspection and grading "as on-line (in-plant) or lot inspection...upon application, on a fee-for-service basis.") . Honey is graded based upon a number of factors, including water content, flavor and aroma, absence of defects and clarity. Honey is also classified by color though it is not a factor in the grading scale. U.S. honey grade scales are Grade A, Grade B, Grade C and Grade substandard.

Lobster

A lobster

In the U.S., lobster grading involves denoting lobsters as new-shell, hard-shell or old-shell, and because lobsters which have recently shed their shells are the most delicate, there is an inverse relationship between the price of American lobster and its flavour. New-shell lobsters have paper-thin shells and a worse meat-to-shell ratio, but the meat is very sweet. However, the lobsters are so delicate that even transport to Boston almost kills them, making the market for new-shell lobsters strictly local to the fishing towns where they are offloaded. Hard-shell lobsters with firm shells, but with less sweet meat, can survive shipping to Boston, New York and even Los Angeles, so they command a higher price than new-shell lobsters. Meanwhile, old-shell lobsters, which have not shed since the previous season and have a coarser flavour, can be air-shipped anywhere in the world and arrive alive, making them the most expensive. One seafood guide notes that an eight-dollar lobster dinner at a restaurant overlooking fishing piers in Maine is consistently delicious, while "the eighty-dollar lobster in a three-star Paris restaurant is apt to be as much about presentation as flavor".

Maple Syrup

Maple syrup from Quebec, graded as No. 1, Light (enlarge image to read the label atop the can)

In Canada, the grading of maple syrup involves classification by the Canadian Food Inspection Agency (CFIA) of maple syrups into one of three grades, each with several colour classes: Canada No. 1, including Extra Light, Light, and Medium; No. 2 Amber; and finally No. 3 Dark or any other ungraded category. Producers in Quebec and Ontario may follow either federal or provincial grading guidelines, which differ slightly.

The United States uses different grading standards. Maple syrup is divided into two major grades: Grade A and Grade B. Grade A is further divided into three subgrades: Light Amber (sometimes known as Fancy), Medium Amber, and Dark Amber. The Vermont Agency of Agriculture Food and Markets uses a similar grading system of color, and is roughly equivalent, especially for lighter syrups, but using letters: "AA", "A", etc. The Vermont grading system differs from the US system in maintaining a slightly higher standard of product density (measured on the Baumé scale). New Hampshire maintains a similar standard, but not a separate state grading scale. The Vermont-graded product has 0.9 percent more sugar and less water in its composition than US-graded. One grade of syrup not for table use, called commercial or Grade C, is also produced under the Vermont system. Vermont inspectors enforce strict syrup grading regulations, and can fine producers up to US$1000 for labelling syrup incorrectly.

Milk

In the United States, there are two grades of milk, with Grade A primarily used for direct sales and consumption in stores, and Grade B used for indirect consumption, such as in cheese making or other processing.

The differences between the two grades are defined in the Wisconsin administrative code for Agriculture, Trade, and Consumer Protection, chapter 60. Grade B generally refers to milk that is cooled in milk cans, which are immersed in a bath of cold flowing water that typically is drawn up from an underground water well rather than using mechanical refrigeration.

Oranges

The USDA has established the following grades for Florida oranges, which primarily apply to oranges sold as fresh fruit: US Fancy, US No. 1 Bright, US No. 1, US No. 1 Golden, US No. 1 Bronze, US No. 1 Russet, US No. 2 Bright, US No. 2, US No. 2 Russet, and US No. 3. The general characteristics graded are color (both hue and uniformity), firmness, maturity, varietal characteristics, texture, and shape. Fancy, the highest grade, requires the highest grade of color and an absence of blemishes, while the terms Bright, Golden, Bronze, and Russet concern solely discoloration.

Florida navel oranges

Peas

Peas

Pea grading involves sorting peas by size, in which smallest peas are graded as the highest quality for their tenderness. Brines may be used, in which peas are floated in them, from which their density can be determined.

Potatoes

In the U.S., potato grading for Idaho potatoes is performed in which No. 1 potatoes are the highest quality and No. 2 are rated as lower in quality due to their appearance (e.g. blemishes or bruises, pointy ends). Density assessment can be performed by floating them in brines. High density potatoes are desirable in the production of dehydrated mashed potatoes, potato crisps and french fries.

Rice

The main criteria used by many countries and millers in rice grading are degree of milling, appearance (color), damaged (broken) and percentage of chalky kernels. In the United States rice is marketed according to three main properties size, color and condition (kernels damage), these properties are directly related to quality, milling percentage and other processing conditions. All properties are considered important in grading. For instance, chalky kernels are not desirable because they give lower milling yields after processing and easily break during handling.

Spices

Cinnamon

In Sri Lanka, cinnamon grading is performed by dividing cinnamon quills into four groups, which are then further divided into specific grades.

Vanilla

Several vanilla fruit grading systems are in use. Each country which produces vanilla has its own grading system, and individual vendors, in turn, sometimes use their own criteria for describing the quality of the fruits they offer for sale.

People grading vanilla beans in Sambava, Madagascar

Tea

In the western black tea industry, tea leaf grading is the process of evaluating products based on the quality and condition of the tea leaves themselves. The highest grades are referred to as "orange pekoe", and the lowest as "fannings" or "dust". This grading system is based upon the size of processed and dried black tea leaves. Despite a purported Chinese origin, these grading terms are typically used for teas from Sri Lanka, India and countries other than China; they are not generally known within Chinese-speaking countries.

Black Tea

Black tea grading is usually based upon one of four scales of quality. Whole-leaf teas are the highest quality, followed by broken leaves, fannings, and dusts. Whole-leaf teas are produced with little or no alteration to the tea leaf. This results in a finished product with a coarser texture than that of bagged teas. Whole-leaf teas are widely considered the most valuable, especially if they contain leaf tips. Broken leaves are commonly sold as medium-grade loose teas. Smaller broken varieties may be included in tea bags.

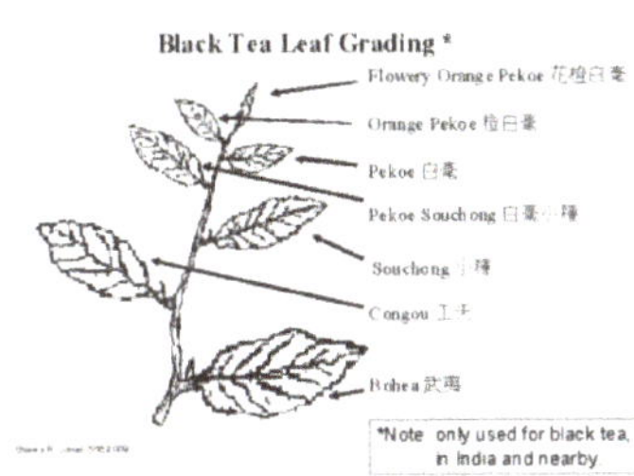

A black tea leaf grading diagram

Fresh tea leaves of different sizes

Rooibos

Rooibos grades are largely related to the proportion of "needle" or leaf to stem content in the mix. A higher leaf content will result in a darker liquor, richer flavour and less "dusty" aftertaste. The high-grade rooibos is exported and does not reach local markets, with major consumers being in the EU, particularly Germany, where it is used in creating flavoured blends for loose-leaf tea markets. In development within South Africa are a small number of specialty tea companies producing similar blends.

Organic green rooibos tea leaves

Food Additives and Preservatives

Carrageenan

There are two basic grades of carrageenan, refined carrageenan (RC) and semi-refined carrageenan (SRC). In the United States, RC and SRC are both labeled as carrageenan. In the European Union, RC is designated by the E number E-407, and SRC is E-407a. RC has a 2% maximum for acid-insoluble material and is produced through an alcohol precipitation process or potassium chloride gel press process. SRC contains a much higher level of cellulosic content and is produced in a less complex process. Indonesia, the Philippines, and Chile are three main sources of raw material and extracted carrageenan.

Lye

Lye is used to cure foods such as lutefisk, olives (making them less bitter), canned mandarin oranges, hominy, lye rolls, century eggs, and pretzels. It is also used as a tenderizer in the crust of baked Cantonese moon cakes, and in lye-water "zongzi" (glutinous rice dumplings wrapped in bamboo leaves), chewy southern Chinese noodles popular in Hong Kong and southern China, and Japanese ramen noodles. In the United States, food-grade lye must meet the requirements outlined in the Food Chemicals Codex (FCC), as prescribed by the U.S. Food and Drug Administration (FDA). Lower or more caustic grades of lye are commonly used as drain de-cloggers and oven cleaners. As a result, such grades should not be used for food preparation, as they would be too poisonous for human consumption.

Sodium Bisulfate

Sodium bisulfate is used as a food additive to leaven cake mixes (make them rise) as well as being used in meat and poultry processing and most recently in browning prevention of fresh-cut produce. The food-grade product meets the requirements set out in the Food Chemicals Codex. It is denoted by E number E514ii in the EU and is approved for use in Australia and New Zealand where it is listed as additive 514. Food-grade sodium bisulfate is used in a variety of food products, including beverages, dressings, sauces, and fillings. It has many synonyms (bisulfate of soda, sodium bisulfate, sodium acid sulfate, monosodium hydrogen sulfate, monosodium salt, sodium hydrogen sulfate, sodium hydrosulfate, sodium pyrosulfate, sulfuric acid, sulfuric acid sodium salt (1:1)).

By Country

- In India, AGMARK is a certification mark employed on agricultural products, assuring that they conform to a set of standards approved by the Directorate of Marketing and Inspection, an agency of the Government of India.

Food Fortification

Manufacturers once proposed selling fortified junk food and beer, but USFDA policies of the time forbade it

Food fortification or enrichment is the process of adding micronutrients (essential trace elements and vitamins) to food. It may be a purely commercial choice to provide extra nutrients in a food, while other times it is a public health policy which aims to reduce the number of people with dietary deficiencies within a population.

Diets that lack variety can be deficient in certain nutrients. Sometimes the staple foods of a region can lack particular nutrients, due to the soil of the region or because of the inherent inadequacy of the normal diet. Addition of micronutrients to staples and condiments can prevent large-scale deficiency diseases in these cases.

While it is true that both fortification and enrichment refer to the addition of nutrients to food, the true definitions do slightly vary. As defined by the World Health Organization (WHO) and the Food and Agricultural Organization of the United Nations (FAO), fortification refers to "the practice of deliberately increasing the content of an essential micronutrient, ie. vitamins and minerals (including trace elements) in a food irrespective of whether the nutrients were originally in the food before processing or not, so as to improve the nutritional quality of the food supply and to provide a public health benefit with minimal risk to health," whereas enrichment is defined as "synonymous with fortification and refers to the addition of micronutrients to a food which are lost during processing."

Food fortification was identified as the second strategy of four by the WHO and FAO to begin decreasing the incidence of nutrient deficiencies at the global level.

As outlined by the FAO, the most common fortified foods are:

- Cereals and cereal based products
- Milk and Milk products
- Fats and oils
- Accessory food items
- Tea and other beverages
- Infant formulas

Types

The four main methods of food fortification (named as to indicate the procedure that is used in order to fortify the food):

1. Biofortification (i.e. breeding crops to increase their nutritional value, which can include both conventional selective breeding, and modern genetic modification)
2. Synthetic biology (i.e. addition of probiotic bacteria to foods)
3. Commercial and industrial fortification (i.e. flour, rice, oils (common cooking foods))
4. Home fortification (e.g. vitamin D drops)

Rationale

The WHO and FAO, among many other nationally recognized organizations, have recognized that there are over 2 billion people worldwide who suffer from a variety of micronutrient deficiencies. In 1992, 159 countries pledged at the FAO/WHO International Conference on Nutrition to make efforts to help combat these issues of micronutrient deficiencies, highlighting the importance of decreasing the number of those with iodine, vitamin A, and iron deficiencies. A significant statistic that led to these efforts was the discovery that approximately 1 in 3 people worldwide were at risk for either an iodine, vitamin A, or iron deficiency. Although it is recognized that food fortification alone will not combat this deficiency, it is a step towards reducing the prevalence of these deficiencies and their associated health conditions.

In Canada, The Food and Drug Regulations have outlined specific criterion which justifies food fortification:

1. To replace nutrients which were lost during manufacturing of the product (i.e. the manufacturing of flour)
2. To act as a public health intervention
3. To ensure the nutritional equivalence of substitute foods (i.e. to make butter and margarine similar in content, soy milk and cow's milk, etc.)
4. To ensure the appropriate vitamin and mineral nutrient composition of foods for special dietary purposes (i.e.,gluten-free products, low sodium, or any other products specifically designed for special dietary requirements from an individual).

There are also several advantages to approaching nutrient deficiencies among populations via food fortification as opposed to other methods. These may include, but are not limited to: treating a population without specific dietary interventions therefore not requiring a change in dietary patterns, continuous delivery of the nutrient, does not require individual compliance, and potential to maintain nutrient stores more efficiently if consumed on a regular basis.

Criticism

Several organizations such as the WHO, FAO, Health Canada, and the Nestlé Research Center acknowledge that there are limitations to food fortification. Within the discussion of nutrient deficiencies the topic of nutrient toxicities can also be immediately questioned. Fortification of nutrients in foods may deliver toxic amounts of nutrients to an individual and also cause its associated side effects. As seen with the case of excessive fluoride intakes below, the result can be irreversible staining to the teeth. Although this may be a minor toxic effect to health, there are several that are more severe.

The WHO states that limitations to food fortification may include: human rights issues indicating that consumers have the right to choose if they want fortified products or not, the potential for insufficient demand of the fortified product, increased production costs leading to increased retail costs, the potential that the fortified products will still not be a solution to nutrient deficiencies amongst low income populations who may not be able to afford the new product, and children who may not be able to consume adequate amounts thereof.

Food safety worries led to legislation in Denmark in 2004 restricting foods fortified with extra vitamins or minerals. Products banned include: Rice Crispies, Shreddies, Horlicks, Ovaltine and Marmite.

Danes said [Kelloggs] Corn Flakes, Rice Krispies and Special K wanted to include "toxic" doses which, if eaten regularly, could damage children's livers and kidneys and harm fetuses in pregnant women.

Limited Absorption

One factor that limits the benefits of food fortification is that isolated nutrients added back into a processed food that has had many of its nutrients removed, does not always result in the added nutrients being as bioavailable as they would be in the original, whole food. An example is skim milk that has had the fat removed, and then had vitamin A and vitamin D added back. Vitamins A and D are both fat-soluble and non-water-soluble, so a person consuming skim milk in the absence of fats may not be able to absorb as much of these vitamins as one would be able to absorb from drinking whole milk. On the other hand, the nutrient added as a fortificant may have a higher bioavailability than from foods, which is the case with folic acid used to increase folate intakes.

Phytochemicals such as polyphenols, phytic acid in cereal grains and tannin in tea can also impact nutrient absorption, limiting the bioavailability of intrinsic and additional nutrients, and reducing the effectiveness of fortification programs.

Excess Intake of Micronutrients

Ecological studies have shown that increased B vitamin fortification is correlated with the prevalence of obesity and diabetes. Daily consumption of iron per capita in the United States has dramatically surged since World War II and nearly doubled over the past century due to increases in iron fortification and increased consumption of meat. Existing evidence suggests that excess iron intake may play a role in the development of obesity, cardiovascular disease, diabetes and cancer.

Fortification of foods with folic acid has been mandated in many countries solely to improve the folate status of pregnant women to prevent Neural Tube Defects—a relatively rare birth defect which affected 0.5% of US births before fortification began. However, when fortification is introduced, several hundred thousand people are exposed to an increased intake of folic acid for each neural tube defect pregnancy that is prevented. In humans, increased folic acid intake leads to elevated blood concentrations of naturally occurring folates and of unmetabolized folic acid. High blood concentrations of folic acid may decrease natural killer cell cytotoxicity, and high folate status may reduce the response to drugs used to treat malaria, rheumatoid arthritis, psoriasis, and cancer. A combination of high folate levels and low vitamin B-12 status may be associated with an increased risk of cognitive impairment and anemia in the elderly and, in pregnant women, with an increased risk of insulin resistance and obesity in their children. Folate has a dual effect on cancer, protecting against cancer initiation but facilitating progression and growth of preneoplastic cells and subclinical cancers. Furthermore, intake of folic acid from fortification have turned out to be significantly greater than originally modeled in pre mandate predictions. Therefore, a high folic acid intake due to fortification may be harmful for more people than the policy is designed to help.

Different Forms of Micronutrients

There is a concern that micronutrients are legally defined in such a way that does not distinguish between different forms, and that fortified foods often have nutrients in a balance that would not occur naturally. For example, in the U.S., food is fortified with folic acid, which is one of the many naturally-occurring forms of folate, and which only contributes a minor amount to the folates occurring in natural foods. In many cases, such as with folate, it is an open question of whether or not there are any benefits or risks to consuming folic acid in this form.

In many cases, the micronutrients added to foods in fortification are synthetic.

In some cases, certain forms of micronutrients can be actively toxic in a sufficiently high dose, even if other forms are safe at the same or much higher doses. There are examples of such toxicity in both synthetic and naturally-occurring forms of vitamins. Retinol, the active form of Vitamin A, is toxic in a much lower dose than other forms, such as beta carotene. Menadione, a phased-out synthetic form of Vitamin K, is also known to be toxic.

Food Supplements

There are several main groups of food supplements like:

- Vitamins and co-vitamins
- Essential minerals
- Essential fatty acids
- Essential amino acids
- Phytonutrients
- Enzymes

Examples of Fortification in Foods

Many foods and beverages worldwide have been fortified, whether a voluntary action by the product developers or by law. Although some may view these additions as strategic marketing schemes to sell their product, there is a lot of work that must go into a product before simply fortifying it. In order to fortify a product, it must first be proven that the addition of this vitamin or mineral is beneficial to health, safe, and an effective method of delivery. The addition must also abide by all food and labeling regulations and support nutritional rationale. From a food developer's point of view, they also need to consider the costs associated with this new product and whether or not there will be a market to support the change.

Examples of foods and beverages that have been fortified and shown to have positive health effects:

Iodized Salt

"Iodine deficiency disorder (IDD) is the single greatest cause of preventable mental retardation. Severe deficiencies cause cretinism, stillbirth and miscarriage. But even mild deficiency can significantly affect the learning ability of populations........ Today over 1 billion people in the world suffer from iodine deficiency, and 38 million babies born every year are not protected from brain damage due to IDD."—Kul Gautam, Deputy Executive Director, UNICEF, October 2007

Iodised salt has been used in the United States since before World War II. It was discovered in 1821 that goiters could be treated by the use of iodized salts. However, it was not until 1916 that the use of iodized salts could be tested in a research trial as a preventative measure against goiters. By 1924, it became readily available in the US.

Currently in Canada and the US, the RDA for iodine is as low as 90 μg/day for children (4–8 years) and as high as 290 μg/day for breast-feeding mothers.

Diseases that are associated with an iodine deficiency include: mental retardation, hypothyroidism, and goiter. There is also a risk of various other growth and developmental abnormalities.

Folic Acid

Folic acid (also known as folate) functions in reducing blood homocysteine levels, forming red blood cells, proper growth and division of cells, and preventing neural tube defects (NTDs).

In many industrialized countries, the addition of folic acid to flour has prevented a significant number of NTDs in infants. Two common types of NTDs, spina bifida and anencephaly, affect approximately 2500-3000 infants born in the US annually. Research trials have shown the ability to reduce the incidence of NTDs by supplementing pregnant mothers with folic acid by 72%.

The RDA for folic acid ranges from as low as 150 μg/day for children aged 1–3 years old, to 400 μg/day for males and females over the age of 19, and 600 μg/day during pregnancy.

Diseases associated with folic acid deficiency include: megaloblastic or macrocytic anemia, cardiovascular disease, certain types of cancer, and NTDs in infants.

Niacin

Niacin has been added to bread in the USA since 1938 (when voluntary addition started), a programme which substantially reduced the incidence of pellagra. As early as 1755, pellagra was recognized by doctors as being a niacin deficiency disease. Although not officially receiving its name of pellagra until 1771. Pellagra was seen amongst poor families who used corn as their main dietary staple. Although corn itself does contain niacin, it is not a bioavailable form unless it undergoes Nixtamalization (treatment with alkali, traditional in Native American cultures) and therefore was not contributing to the overall intake of niacin. Although pellagra can still be seen in developing countries, fortification of food with niacin played a huge role in eliminating the prevalence of the disease.

The RDA for niacin is 2 mg NE(niacin equivalents)/day (AI) for infants aged 0–6 months, 16 mg NE/day for males, and 14 mg NE/day for females who are over the age of 19.

Diseases associated with niacin deficiency include: Pellagra which consisted of signs and symptoms called the 3D's-"Dermatitis, dementia, and diarrhea. Others may include vascular or gastrointestinal diseases.

Common diseases which present a high frequency of niacin deficiency: alcoholism, anorexia nervosa, HIV infection, gastrectomy, malabsorptive disorders, certain cancers and their associated treatments.

Vitamin D

Since Vitamin D is a fat-soluble vitamin, it cannot be added to a wide variety of foods. Foods that it is commonly added to are margarine, vegetable oils and dairy products. During the late 1800s, after the discovery of curing conditions of scurvy and beriberi had occurred, researchers were aiming to see if the disease, later known as rickets, could also be cured by food. Their results showed that sunlight exposure and cod liver oil were the cure. It was not until the 1930s that vitamin D was actually linked to curing rickets. This discovery led to the fortification of common foods such as milk, margarine, and breakfast cereals. This took the astonishing statistics of approximately 80–90% of children showing varying degrees of bone deformations due to vitamin D deficiency to being a very rare condition.

Risk factors for vitamin D deficiencies include:

- In infants, being exclusively or primarily breast-fed
- Dark skin
- Living in cold climates and having little sun exposure
- Being elderly
- Covering all or almost all of one's skin while outdoors
- Liberal use of high-SPF sunscreens
- Fat malabsorption syndromes
- Inflammatory bowel diseases
- Obesity

The current RDA for infants aged 0–6 months is 10 μg (400 International Units (IU))/day and for adults over 19 years of age it is 15 μg (600 IU)/day.

Diseases associated with a vitamin D deficiency include rickets, osteoporosis, and certain types of cancer (breast, prostate, colon and ovaries). It has also been associated with increased risks for fractures, heart disease, type 2 diabetes, autoimmune and infectious diseases, asthma and other wheezing disorders, myocardial infarction, hypertension, congestive heart failure, and peripheral vascular disease.

Fluoride

Although fluoride is not considered an essential mineral, it is seen as possibly useful in prevention of tooth decay and maintaining adequate dental health. In the mid-1900s it was discovered that towns with a high level of fluoride in their water supply was causing the residents' teeth to have both brown spotting and a strange resistance to dental caries. This led to the fortification of water supplies with fluoride with safe amounts to retain the properties of resistance to dental caries but avoid the staining cause by fluorosis (a condition caused by excessive fluoride intake). The tolerable upper intake level (UL) set for fluoride ranges from 0.7 mg/day for infants aged 0–6 months and 10 mg/day for adults over the age of 19.

Conditions commonly associated with fluoride deficiency are dental caries and osteoporosis.

Others

Some other examples of fortified foods:

- Calcium is frequently added to fruit juices, carbonated beverages and rice.
- White rice is frequently enriched to replace some of the lost nutrients during milling or adding extras in.
- "Golden rice" is a variety of rice which has been genetically modified to produce beta carotene.
- Amylase rich flour is utilized for food making to increase dietary consumption.

Fortification for Body Building

Despite having some scientific basis, but with controversial ethics, is the science of using foods and food supplements to achieve a defined health goal. A common example of this use of food supplements is the extent to which body builders will use amino acid mixtures, vitamins and phytochemicals to enhance natural hormone production, increase muscle and reduce fat. The literature is not concrete on an appropriate method for use of fortification for body builders and therefore may not be recommended due to safety concerns.

Fortification for Medical Treatment

There is interest in the use of food supplements in established medical conditions. This nutritional supplementation using foods as medicine (nutraceuticals) has been effectively used in treating disorders affecting the immune system up to and including cancers. This goes beyond the definition of "food supplement", but should be included for the sake of completeness.

Food Sampling

a food scientist considers canned peas

Food sampling is a process used to check that a food is safe and that it does not contain harmful contaminants, or that it contains only permitted additives at acceptable levels, or that it contains the right levels of key ingredients and its label declarations are correct, or to know the levels of nutrients present.

A food sample is carried out by subjecting the product to physical analysis. Analysis may be undertaken by or on behalf of a manufacturer regarding their own product, or for official food law enforcement or control purposes, or for research or public information.

To undertake any analysis, unless the whole amount of food to be considered is very small so that the food can be used for testing in its entirety, it is usually necessary for a portion of it to be taken (e.g. a small quantity from a full production batch, or a portion of what is on sale in a shop) – this process is known as food sampling.

In most cases with food to be analysed there are two levels of sampling – the first being selection of a portion from the whole, which is then submitted to a laboratory for testing, and the second being the laboratory's taking of the individual amounts necessary for individual tests that may be applied. It is the former that is 'food sampling': the latter is analytical laboratory 'sub-sampling', often relying upon initial homogenisation of the entire submitted sample.

Where it is intended that the results of any analysis to relate to the food as a whole it is crucially important that the sample is representative of that whole – and the results of any analysis can only be meaningful if the sampling is undertaken effectively. This is true whether the 'whole' is a manufacturer's entire production batch, or where it is a single item but too large to all be used for the test.

Factors relevant in considering the representativeness of a sample include the homogeneity of the food, the relative sizes of the sample to be taken and the whole, the potential degree of variation of the parameter(s) in question through the whole, and the significance and intended use of the analytical result.

Sampling by Manufacturers

Food manufacturers and producers would need to satisfy themselves that any sample taken for analysis is sufficiently representative of the food for the analytical result to be meaningful. This is true whether the data are to be used as the basis of labelling declarations, assurance of compliance with legislative or other standards, monitoring of production as part of HACCP (Hazard Analysis and Critical Control Points), or for routine quality control.

In the United Kingdom although various guidance is available, either from manufacturers' associations or from sources of standards such as British Standards Institution (such as British Standard BS6001), some of which may be relevant to certain food types. It is largely down to manufacturers to make their own evaluations of need and suitability. This must be translated into an assessment both of sample portion size and number, and the frequency of taking samples.

Testing and Methods of Analysis

To ensure food safety and quality, some food samples which are perishable require certain tests and analyses. The following tests and analyses can be conducted:

- Food allergen testing
- Food chemical analysis
- Food contact tests

- Food contaminant testing
- Nutritional analysis and testing
- GMO testing
- Melamine contamination testing
- Microbiological tests
- Spiral plating for bacterial count
- Pesticide residue testing
- Veterinary drug residue testing
- PCR food testing

Food Law Enforcement (UK)

In the United Kingdom, enforcement is under the Food Safety Act 1990. Food sampling is undertaken primarily by local authorities and port health authorities for submission to public analysts for analysis. Much of the legislation relates to food as supplied to a consumer, meaning that every portion of a size of perishable food and foods at risk as may be supplied to a consumer has to comply, so that in such cases the sample submitted for analysis could simply be an entire consumer-sized portion. There are exceptions, however, such as the sampling of nut products for the presence of aflatoxins, which stipulate a primary sample size related to the size of the consignment – with associated requirements for initial homogenisation to produce a smaller sample to be sent for analysis.

The Food Safety Act 1990 affords a right for defence analysis, and for referee analysis in case of disputed analytical results, by stipulating that except where to do so would prevent effective analysis the sample must be divided into three parts. The UK Food Standards Agency provides supplementary guidance to the enforcement authorities to assist with the sampling process and associated decisions by sampling officers.

There is no set frequency or rate for the sampling of food for law enforcement in the UK. Between the 1930s and 1990s there had been a guideline minimum rate for sampling for chemical analysis (not including samples for microbiological examination) of 2.5 samples per annum per 1000 head of population, however that was an arbitrary figure and more recent thinking suggested that the selection of a frequency for sampling should be based on risk. In this context risk includes all 'consumer protection' issues such as pecuniary disadvantage from substandard or counterfeit products, as well as risk to health. The Association of Public Analysts was commissioned by the Food Standards Agency to look into this, culminating in a scheme for *Risk Based Sampling*, though it has not yet been adopted by the enforcement authorities.

Optical Sorting

Optical sorting (sometimes called digital sorting) is the automated process of sorting solid products using cameras and/or lasers.

Depending on the types of sensors used and the software-driven intelligence of the image processing system, optical sorters can recognize objects' color, size, shape, structural properties and chemical composition. The sorter compares objects to user-defined accept/reject criteria to identify and remove defective products and foreign material (FM) from the production line, or to separate product of different grades or types of materials.

Optical sorting achieves non-destructive, 100 percent inspection in-line at full production volumes.

Optical sorters are in widespread use in the food industry worldwide, with the highest adoption in processing harvested foods such as potatoes, fruits, vegetables and nuts where it achieves non-destructive, 100 percent inspection in-line at full production volumes. The technology is also used in pharmaceutical manufacturing and nutraceutical manufacturing, tobacco processing, waste recycling and other industries. Compared to manual sorting, which is subjective and inconsistent, optical sorting helps improve product quality, maximize throughput and increase yields while reducing labor costs.

The Sorting System

In general, optical sorters feature four major components: the feed system, the optical system, image processing software and the separation system. The objective of the feed system is to spread product into a uniform monolayer so products are presented to the optical system evenly, without clumps, at a constant velocity. The optical system includes lights and sensors housed above and/or below the flow of the objects being inspected. The image processing system compares objects to user-defined accept/reject thresholds to classify objects and actuate the separation system. The separation system, usually compressed air for small products and mechanical devices for larger products like whole potatoes, pinpoints objects while in-air and deflects the objects to remove into a reject chute while good product continues along its normal trajectory.

The ideal sorter depends on the application, with the product's characteristics and the user's objectives determining the ideal sensors, software-driven capabilities and mechanical platform.

Sensors

Optical sorters require a compatible combination of light and sensors to illuminate objects and capture images of the objects before the images can be processed and accept/reject decisions made.

There are camera sorters, laser sorters and sorters that feature a combination of cameras and lasers on one platform. Lights, cameras, lasers and laser sensors can be designed to function within visible light wavelengths as well as the infrared (IR) and ultraviolet (UV) spectrums. The optimal wavelengths for each application maximize the contrast between the objects to be separated. Cameras and laser sensors can differ in spatial resolution, with higher resolutions enabling the sorter to detect and remove smaller defects.

Cameras

Shape sorting enables the detection of same-color defects and foreign material

Monochromatic cameras detect shades of gray from black to white and can be effective when sorting products with high-contrast defects.

Sophisticated color cameras with high color resolution are capable of detecting millions of colors to better distinguish more subtle color defects. Trichromatic color cameras (also called three-channel cameras) divide light into three bands, which can include red, green and/or blue within the visible spectrum as well as IR and UV.

Coupled with intelligent software, sorters that feature cameras are capable of recognizing each object's color, size and shape as well as the color, size, shape and location of a defect on a product. Some intelligent sorters even allow the user to define a defective product based on the total defective surface area of any given object.

Lasers

While cameras capture product information based primarily on material reflectance, lasers and their sensors are able to interrogate a material's structural properties in addition to determining differences in color. This structural property inspection capability makes lasers ideal for detecting a wide range of organic and inorganic foreign material such as insects, glass, metal, sticks, rocks and plastic, even if they are the same color as the good product, and for separating various materials at waste recycling facilities.

Lasers can be designed to operate within specific wavelengths of light, within the visible spectrums and beyond. For example, lasers can detect chlorophyll by stimulating fluorescence using specific wavelengths, a process that is very effective for removing foreign material from green vegetables.

Camera/Laser Combinations

Sorters equipped with cameras and lasers on one platform are generally capable of identifying the widest variety of attributes. Cameras are often better at recognizing color, size and shape while laser sensors identify differences in structural properties to maximize foreign material detection and removal.

Hyperspectral Imaging

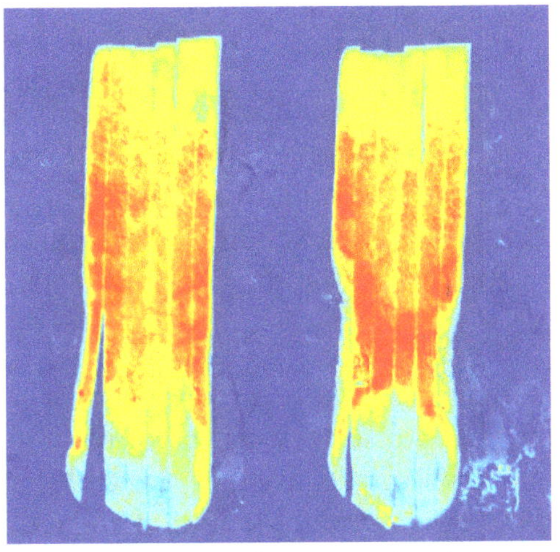

Hyperspectral image of "sugar end" potato strips shows invisible defects

Driven by the need to solve previously impossible sorting challenges, a new generation of sorters that feature multispectral and hyperspectral imaging systems are being developed.

Like trichromatic cameras, multispectral and hyperspectral cameras collect data from across the electromagnetic spectrum. Unlike trichromatic cameras, which divide light into three bands, hyperspectral systems can divide light into hundreds of narrow bands over a continuous range that covers a vast portion of the electromagnetic spectrum. Compared to the three data points per pixel collected by trichromatic cameras, hyperspectral cameras can collect hundreds of data points per pixel, which are combined to create a unique spectral signature (also called a fingerprint) for each object. When complemented by capable software intelligence, a hyperspectral sorter processes those fingerprints to enable sorting on the chemical composition of the product. This is an emerging area of chemometrics.

Software-driven Intelligence

Once the sensors capture the object's response to the energy source, image processing takes over to manipulate the raw data to extract and categorize information about specific features. As raw data flows from the sensors, the definitions of good and bad flow from the user who sets the accept/reject thresholds. The art and science of image processing lies in developing algorithms that maximize the effectiveness of the sorter while presenting a simple user-interface to the operator.

Object-based recognition is a classic example of software-driven intelligence. It allows the user to define a defective product based on where a defect lies on the product and/or the total defective surface area of an object. It offers more control in defining a wider range of defective products, and if used to control the sorter's ejection system, it improves the accuracy of ejecting defective products, which improves product quality and increases yields.

New software-driven capabilities are constantly being developed to address the specific needs of various applications. As computing hardware becomes more powerful, new software-driven advancements that require more bandwidth become possible. Some of these advancements enhance the effectiveness of sorters to achieve better results while others enable completely new sorting decisions to be made.

Platforms

The considerations that determine the ideal platform for a specific application include the nature of the product – large or small, wet or dry, fragile or unbreakable, round or easy to stabilize – and the user's objectives. In general, products smaller than a grain of rice and as large as whole potatoes can be sorted. Throughputs range from less than 2 metric tons of product per hour on low-capacity sorters to more than 35 metric tons of product per hour on high-capacity sorters.

Channel Sorters

The simplest optical sorters are channel sorters, a type of color sorter that can be effective for products that are small, hard and dry with a consistent size and shape such as rice and seeds. For these products, channel sorters offer an affordable solution and ease of use with a small footprint. Channel sorters feature monochromatic or color cameras and remove defects and foreign material based only on differences in color.

For products that are soft or wet or non-homogenous, which cannot be handled by a channel sorter, and for processors that want more control over the quality of their product, freefall sorters (also called waterfall or gravity-fed sorters), chute-fed sorters or belt sorters are ideal. These more sophisticated sorters often feature advanced cameras and/or lasers that, when complemented by capable software intelligence, detect objects' size, shape, color, structural properties and chemical composition.

Freefall and Chute-fed Sorters

As the names imply, freefall sorters inspect product in-air during the freefall, while chute-fed sorters stabilize product on a chute prior to in-air inspection. The major advantages of freefall and chute-fed sorters, compared to belt sorters, are a lower price point, smaller footprint and the absence of moving parts, which contributes to low maintenance. These sorters are often most suitable for nuts and berries as well as a frozen and dried fruits, vegetables, potato strips and seafood, in addition to waste recycling applications that require mid-volume throughputs.

Belt Sorters

Belt sorting platforms are often preferred for higher capacity applications such as vegetable and potato products prior to canning, freezing or drying, most fresh cut produce and wet fruits as well as waste recycling. Belt sorters stabilize product on a conveyor belt prior to inspection. Some belt sorters inspect product on the belt from the top only while others also launch product off the belt for in-air inspection from the bottom. Belt sorters can be designed to achieve traditional two-way sorting or can be equipped with two ejector systems and three outfeed streams to achieve three-way sorting.

Optical sorters can function within visible light wavelengths as well as the IR and UV spectrums

ADR Systems

A fifth type of sorting platform, called an automated defect removal (ADR) system, is specifically for potato strips (French fries). Unlike other sorters that eject products with defects from the production line, ADR systems identify defects and actually cut the defects from the strips. In conjunction with a mechanical nubbin grader that follows the ADR, this combination is essentially another type of optical sorting system because it uses optical sensors to identify and remove defects.

Single-file Inspection Systems

The platforms described above all operate in bulk mode, without needing to single-file product prior to inspection. In contrast, a sixth type of platform, used in the pharmaceutical industry, is a single-file optical inspection system. Although these sorters are effective in detecting and removing foreign tablets and capsules and defects based on differences in size, shape and color, they have not been widely adopted due to the high capital costs, low throughput and slow changeover, compared to the belt sorters for tablets, capsules and softgels.

Mechanical Graders

For products that require sorting only by size, mechanical grading systems that don't utilize sensors or image processing systems are often highly effective. These mechanical grading systems are sometimes referred to as sorting systems, but should not be confused with optical sorters that feature sensors and image processing systems.

Food Intolerance

Food intolerance (or non-allergic food hypersensitivity) is a detrimental reaction, often delayed, to a food, beverage, food additive, or compound found in foods that produces symptoms in one or more body organs and systems, but generally refers to reactions other than food allergy. Food allergies are immune reactions, involving mast cells in which food triggers release of mast cell mediators, such as histamine when the food allergen binds to pre-formed IgE antibodies already bound to the mast cell. This mechanism causes allergies to typically give immediate reaction (a few minutes to a few hours) to foods. There are numerous causes for other types of food sensitivities, most of which, in contrast to food allergies, cause delayed hypersensitivity reactions to the food.

Food intolerances can be classified according to their mechanism. Intolerance can result from the absence of specific chemicals or enzymes needed to digest a food substance, as in hereditary fructose intolerance. It may be a result of an abnormality in the body's ability to absorb nutrients, as occurs in fructose malabsorption. Food intolerance reactions can occur to naturally occurring chemicals in foods, as in salicylate sensitivity. Drugs sourced from plants, such as aspirin, can also cause these kinds of reactions. Finally, it may be the result of non-IgE-mediated immune responses.

Definitions

Non-allergic food hypersensitivity is the medical name for food intolerance, loosely referred to as *food hypersensitivity*, or previously as *pseudo-allergic reactions*. Non-allergic food hypersensitivity should not be confused with true food allergies.

Food intolerance reactions can include pharmacologic, metabolic, and gastro-intestinal responses to foods or food compounds. Food intolerance does not include either psychological responses or foodborne illness.

A non-allergic food hypersensitivity is an abnormal physiological response. It can be difficult to determine the poorly tolerated substance as reactions can be delayed, dose-dependent, and a particular reaction-causing compound may be found in many foods.

- Metabolic food reactions are due to inborn or acquired errors of metabolism of nutrients, such as in diabetes mellitus, lactase deficiency, phenylketonuria and favism.
- Pharmacological reactions are generally due to low-molecular-weight chemicals which occur either as natural compounds, such as salicylates and amines, or to food additives, such as preservatives, colouring, emulsifiers and taste enhancers. These chemicals are capable of causing drug-like (biochemical) side effects in susceptible individuals.
- Gastro-intestinal reactions can be due to malabsorption or other GI Tract abnormalities.
- Immunological responses are mediated by non-IgE immunoglobulins, where the immune system recognises a particular food as a foreign body.
- Toxins may either be present naturally in food, be released by bacteria, or be due to contamination of food products. Toxic food reactions are caused by the direct action of a food or substance without immune involvement.
- Psychological reactions involve manifestation of clinical symptoms caused not by food but by emotions associated with food. These symptoms do not occur when the food is given in an unrecognisable form.

Elimination diets are useful to assist in the diagnosis of food intolerance. There are specific diagnostic tests for certain food intolerances.

Signs and Symptoms

Non-IgE-mediated food hypersensitivity (food intolerance) is more chronic, less acute, less obvious in its presentation, and often more difficult to diagnose than a food allergy. Symptoms of food intolerance vary greatly, and can be mistaken for the symptoms of a food allergy. While true allergies are associated with fast-acting immunoglobulin IgE responses, it can be difficult to determine the offending food causing a food intolerance because the response generally takes place over a prolonged period of time. Thus, the causative agent and the response are separated in time, and may not be obviously related. Food intolerance symptoms usually begin about half an hour after eating or drinking the food in question, but sometimes symptoms may be delayed by up to 48 hours.

Food intolerance can present with symptoms affecting the skin, respiratory tract, gastrointestinal tract (GIT) either individually or in combination. On the skin may include skin rashes, urticaria (hives), angioedema, dermatitis, and eczema. Respiratory tract symptoms can include nasal congestion, sinusitis, pharyngeal irritations, asthma and an unproductive cough. GIT symptoms include mouth ulcers, abdominal cramp, nausea, gas, intermittent diarrhea, constipation, irritable bowel syndrome, and may include anaphylaxis.

Food intolerance has been found associated with; irritable bowel syndrome and inflammatory bowel disease, chronic constipation, chronic hepatitis C infection, eczema, NSAID intolerance, respiratory complaints, including asthma, rhinitis and headache, functional dyspepsia, eosinophilic esophagitis and ENT illnesses.

Causes

Reactions to chemical components of the diet are more common than true food allergies. They are caused by various organic chemicals occurring naturally in a wide variety of foods, both of animal and vegetable origin more often than to food additives, preservatives, colourings and flavourings, such as sulfites or dyes. Both natural and artificial ingredients may cause adverse reactions in sensitive people if consumed in sufficient amount, the degree of sensitivity varying between individuals.

Pharmacological responses to naturally occurring compounds in food, or chemical intolerance, can occur in individuals from both allergic and non-allergic family backgrounds. Symptoms may begin at any age, and may develop quickly or slowly. Triggers may range from a viral infection or illness to environmental chemical exposure. It occurs more commonly in women, which may be because of hormone differences, as many food chemicals mimic hormones.

A deficiency in digestive enzymes can also cause some types of food intolerances. Lactose intolerance is a result of the body not producing sufficient lactase to digest the lactose in milk; dairy foods which are lower in lactose, such as cheese, are less likely to trigger a reaction in this case. Another carbohydrate intolerance caused by enzyme deficiency is hereditary fructose intolerance.

Celiac disease, an autoimmune disorder caused by an immune response to the protein gluten, results in gluten intolerance and can lead to temporary lactose intolerance.

The most widely distributed naturally occurring food chemical capable of provoking reactions is salicylate, although tartrazine and benzoic acid are well recognised in susceptible individuals. Benzoates and salicylates occur naturally in many foods, including fruits, juices, vegetables, spices, herbs, nuts, tea, wines, and coffee. Salicylate sensitivity causes reactions to not only aspirin and NSAIDs but also foods in which salicylates naturally occur, such as cherries.

Other natural chemicals which commonly cause reactions and cross reactivity include amines, nitrates, sulphites and some antioxidants. Chemicals involved in aroma and flavour are often suspect.

The classification or avoidance of foods based on botanical families bears no relationship to their chemical content and is not relevant in the management of food intolerance.

Salicylate-containing foods include apples, citrus fruits, strawberries, tomatoes, and wine, while reactions to chocolate, cheese, bananas, avocado, tomato or wine point to amines as the likely food chemical. Thus, exclusion of single foods does not necessarily identify the chemical responsible as several chemicals can be present in a food, the patient may be sensitive to multiple food chemicals and reaction more likely to occur when foods containing the triggering substance are eaten in a combined quantity that exceeds the patient's sensitivity thresholds. People with food sensitivities have different sensitivity thresholds, and so more sensitive people will react to much smaller amounts of the substance.

Pathogenesis

The term food allergy is widely misused for all adverse reactions to food. Food allergy (FA) is a food hypersensitivity occurring in susceptible individuals, which is mediated by a classical immune mechanism specific for the food itself. The best established mechanism in FA is due to the presence of IgE antibodies against the offending food. Food intolerance (FI) are all other adverse reactions to food. Subgroups of FI are enzymatic (e.g. lactose intolerance due to lactase deficiency), pharmacological (e.g. reactions against biogenic amines, histamine intolerance), and undefined food intolerance (e.g. against some food additives). As knowledge of mechanisms and causes of food intolerance improve, nomenclature will be updated. There is no worldwide scientific consensus on the pathogenesis of food intolerance.

Food intolerances can be caused by enzymatic defects in the digestive system, can also result from pharmacological effects of vasoactive amines present in foods (e.g. Histamine), among other metabolic, pharmacological and digestive abnormalities.

A frequent misconception among the public is confusion between cow's milk allergy (CMA) and cow's milk intolerance, which is usually intolerance to lactose. There are at least two, and possibly more, distinct pathologies. Hypersensitivity to milk is often broadly classified into immunoglobulin E (IgE)-mediated allergy and non-IgE-mediated intolerance. The immunopathological mechanisms of non-IgE-mediated intolerance in particular remain poorly understood, and this has hindered the development of simple and reliable diagnostics. Adults with non-IgE-mediated intolerance to milk tend to suffer ongoing reactions without the development of tolerance. The precise immunopathological mechanisms of non-IgE-mediated intolerance remain unclear. A number of mechanisms have been implicated, including type-1

T helper cell (Th1) mediated reactions, the formation of immune complexes leading to the activation of Complement, or T-cell/mast cell/neuron interactions inducing functional changes in smooth muscle action and intestinal motility. Food antigens contact the immune system throughout the intestinal tract via the gut associated lymphoid tissue (GALT), where interactions between antigen presenting cells and T cells direct the type of immune response mounted. Unresponsiveness of the immune system to dietary antigens is termed "oral tolerance" and is believed to involve the deletion or switching off of reactive antigen-specific T cells and the production of regulatory T cells (T reg) that quell inflammatory responses to benign antigens. In the case of IgE-mediated allergies, a deficiency in regulation and a polarisation of specific effector T cells towards type-2 T helper cells (Th2) lead to signalling of B-cells to produce milk protein-specific IgE. Whereas non-IgE-mediated reactions (intolerances) may be due to Th1 mediated inflammation. Dysfunctional T reg cell activity has been identified as a factor in both allergy/ intolerance mechanisms.

Diagnosis

Diagnosis of food intolerance can include hydrogen breath testing for lactose intolerance and fructose malabsorption, professionally supervised elimination diets, and ELISA testing for IgG-mediated immune responses to specific foods. It is important to be able to distinguish between food allergy, food intolerance, and autoimmune disease in the management of these disorders. Non-IgE-mediated intolerance is more chronic, less acute, less obvious in its clinical presentation, and often more difficult to diagnose than allergy, as skin tests and standard immunological studies are not helpful. Elimination diets must remove all poorly tolerated foods, or all foods containing offending compounds. Clinical investigation is generally undertaken only for more serious cases, as for minor complaints which do not significantly limit the person's lifestyle the cure may be more inconvenient than the problem.

The Hemocode Food Intolerance System and Rocky Mountain Analytical Food Allergy Test are unvalidated yet heavily marketed examples of ELISA testing of IgG4 to foods. IgG4 against foods indicates that the person has been repeatedly exposed to food proteins recognized as foreign by the immune system. However, its presence should not be considered a factor which induces intolerance. Food-specific IgG4 does not indicate food allergy or intolerance, but rather a normal physiological response of the immune system after exposure to food components. Although elimination of foods based on IgG-4 testing in IBS patients resulted in an improvement in symptoms, the positive effects of food elimination were more likely due to wheat and milk elimination than IgG-4 test-determined factors. The IgG-4 test specificity is questionable as healthy individuals with no symptoms of food intolerance also test positive for IgG-4 to several foods.

Diagnosis is made using medical history and cutaneous and serological tests to exclude other causes, but to obtain final confirmation a Double Blind Controlled Food Challenge must be performed. Treatment can involve long-term avoidance, or if possible re-establishing a level of tolerance.

Today there are many methods available such as Cytotoxic testing, MRT testing, Elisa Testing and Microarray Elisa Testing. Allergy US reviewed these methods and Microarray technology appears to be the most reliable among them. http://allergyus.com/food-intolerance-tests-in-usa/.

Prevention

There is emerging evidence from studies of cord bloods that both sensitization and the acquisition of tolerance can begin in pregnancy, however the window of main danger for sensitization to foods extends prenatally, remaining most critical during early infancy when the immune system and intestinal tract are still maturing. There is no conclusive evidence to support the restriction of dairy intake in the maternal diet during pregnancy in order to prevent. This is generally not recommended since the drawbacks in terms of loss of nutrition can out-weigh the benefits. However, further randomised, controlled trials are required to examine if dietary exclusion by lactating mothers can truly minimize risk to a significant degree and if any reduction in risk is out-weighed by deleterious impacts on maternal nutrition.

A Cochrane review has concluded feeding with a soy formula cannot be recommended for prevention of allergy or food intolerance in infants. Further research may be warranted to determine the role of soy formulas for prevention of allergy or food intolerance in infants unable to be breast fed with a strong family history of allergy or cow's milk protein intolerance. In the case of allergy and celiac disease others recommend a dietary regimen is effective in the prevention of allergic diseases in high-risk infants, particularly in early infancy regarding food allergy and eczema. The most effective dietary regimen is exclusively breastfeeding for at least 4–6 months or, in absence of breast milk, formulas with documented reduced allergenicity for at least the first 4 months, combined with avoidance of solid food and cow's milk for the first 4 months.

Management

Individuals can try minor changes of diet to exclude foods causing obvious reactions, and for many this may be adequate without the need for professional assistance. For reasons mentioned above foods causing problems may not be so obvious since food sensitivities may not be noticed for hours or even days after one has digested food. Persons unable to isolate foods and those more sensitive or with disabling symptoms should seek expert medical and dietitian help. The dietetic department of a teaching hospital is a good start.

Guidance can also be given to your general practitioner to assist in diagnosis and management. Food elimination diets have been designed to exclude food compounds likely to cause reactions and foods commonly causing true allergies and those foods where enzyme deficiency cause symptoms. These elimination diets are not everyday diets but intended to isolate problem foods and chemicals. Avoidance of foods with additives is also essential in this process.

Individuals and practitioners need to be aware that during the elimination process patients can display aspects of food addiction, masking, withdrawals, and further sensitization and intolerance. Those foods that an individual considers as 'must have every day' are suspect addictions, this includes tea, coffee, chocolate and health foods and drinks, as they all contain food chemicals. Individuals are also unlikely to associate foods causing problems because of masking or where separation of time between eating and symptoms occur. The elimination process can overcome addiction and unmask problem foods so that the patients can associate cause and effect.

It takes around five days of total abstinence to unmask a food or chemical, during the first week on an elimination diet withdrawal symptoms can occur but it takes at least two weeks to remove residual traces. If symptoms have not subsided after six weeks, food intolerance is unlikely to be involved and a normal diet should be restarted. Withdrawals are often associated with a lowering of the threshold for sensitivity which assists in challenge testing, but in this period individuals can be ultra-sensitive even to food smells so care must be taken to avoid all exposures.

After two or more weeks if the symptoms have reduced considerably or gone for at least five days then challenge testing can begin. This can be carried out with selected foods containing only one food chemical, to isolate it if reactions occur. In Australia, purified food chemicals in capsule form are available to doctors for patient testing. These are often combined with placebo capsules for control purposes. This type of challenge is more definitive. New challenges should only be given after 48 hours if no reactions occur or after five days of no symptoms if reactions occur.

Once all food chemical sensitivities are identified a dietitian can prescribe an appropriate diet for the individual to avoid foods with those chemicals. Lists of suitable foods are available from various hospitals and patient support groups can give local food brand advice. A dietitian will ensure adequate nutrition is achieved with safe foods and supplements if need be.

Over a period of time it is possible for individuals avoiding food chemicals to build up a level of resistance by regular exposure to small amounts in a controlled way, but care must be taken, the aim being to build up a varied diet with adequate composition.

Prognosis

The prognosis of children diagnosed with intolerance to milk is good: patients respond to diet which excludes cow's milk protein and the majority of patients succeed in forming tolerance. Children with non-IgE-mediated cows milk intolerance have a good prognosis, whereas children with IgE-mediated cows milk allergy in early childhood have a significantly increased risk for persistent allergy, development of other food allergies, asthma and rhinoconjunctivitis.

A study has demonstrated that identifying and appropriately addressing food sensitivity in IBS patients not previously responding to standard therapy results in a sustained clinical improvement and increased overall well being and quality of life.

Epidemiology

Estimates of the prevalence of food intolerance vary widely from 2% to over 20% of the population. So far only three prevalence studies in Dutch and English adults have been based on double-blind, placebo-controlled food challenges. The reported prevalences of food allergy/intolerance (by questionnaires) were 12% to 19%, whereas the confirmed prevalences varied from 0.8% to 2.4%. For intolerance to food additives the prevalence varied between 0.01 and 0.23%.

Food intolerance rates were found to be similar in the population in Norway. Out of 4,622 subjects with adequately filled-in questionnaires, 84 were included in the study (1.8%) Perceived food intolerance is a common problem with significant nutritional consequences in a population with IBS. Of these 59 (70%) had symptoms related to intake of food, 62% limited or excluded food items from the diet. Tests were performed for food allergy and malabsorption, but not for intolerance. There were no associations between the tests for food allergy and malabsorption and perceived food intolerance, among those with IBS. Perceived food intolerance was unrelated to musculoskeletal pain and mood disorders.

According to the RACP working group, "Though not considered a "cause" of CFS, some patients with chronic fatigue report food intolerances that can exacerbate symptoms."

History

In 1978 Australian researchers published details of an 'exclusion diet' to exclude specific food chemicals from the diet of patients. This provided a basis for challenge with these additives and natural chemicals. Using this approach the role played by dietary chemical factors in the pathogenesis of chronic idiopathic urticaria (CIU) was first established and set the stage for future DBPCT trials of such substances in food intolerance studies.

In 1995 the European Academy of Allergology and Clinical Immunology suggested a classification on the basis of the responsible pathogenetic mechanism; according to this classification, non-toxic reactions can be divided into 'food allergies' when they recognize immunological mechanisms, and 'food intolerances' when there are no immunological implications. Reactions secondary to food ingestion are defined generally as 'adverse reactions to food'.

In 2003 the Nomenclature Review Committee of the World Allergy Organization issued a report of revised nomenclature for global use on food allergy and food intolerance, that has had general acceptance. Food intolerance is described as a 'non allergic hypersensitivity' to food.

Society and Culture

In the UK, scepticism about food intolerance as a specific condition influenced doctors' (GPs') perceptions of patients and of the patients' underlying problems. However, rather than risk damaging the doctor-patient relationship, GPs chose - despite their scepticism and guided by an element of awareness of the limitations of modern medicine - to negotiate mutually acceptable ground with patients and with patients' beliefs. As a result, whether due to a placebo effect, a secondary benefit, or a biophysical result of excluding a food from the diet, the GPs acknowledge both personal and therapeutic benefits.

In the Netherlands, patients and their doctors (GPs) have different perceptions of the efficacy of diagnostic and dietary interventions in IBS. Patients consider food intolerance and GPs regard lack of fibre as the main etiologic dietary factor. It has been suggested that Dutch GPs explore the patients' expectations and potentially incorporate these in their approach to IBS patients.

New food labeling regulations were introduced into the USA and Europe in 2006, which are said to benefit people with intolerances. In general, food-allergic consumers were not satisfied with the current labelling practices. In the USA food companies propose distinguishing between food allergy and food intolerance and use a mechanism-based (i.e., immunoglobulin-E-mediated), acute life-threatening anaphylaxis that is standardized and measurable and reflects the severity of health risk, as the principal inclusion criterion for food allergen labeling. Symptoms due to, or exacerbated by, food additives usually involve non-IgE-mediated mechanisms (food intolerance) and are usually less severe than those induced by food allergy, but can include anaphylaxis.

Research Directions

A Low FODMAP diet now has an evidence base sufficiently strong to recommend its widespread application in conditions such as IBS and IBD. Restriction of Fermentable Oligo-, Di- and Monosaccharides and Polyols globally, rather than individually, controls the symptoms of functional gut disorders (e.g. IBS), and the majority of IBD patients respond just as well. It is more successful than restricting only Fructose and Fructans, which are also FODMAPs, as is recommended for those with Fructose malabsorption. Longer term compliance with the diet was high.

A randomised controlled trial on IBS patients found relaxing an IgG-mediated food intolerance diet led to a 24% greater deterioration in symptoms compared to those on the elimination diet and concluded food elimination based on IgG antibodies may be effective in reducing IBS symptoms and is worthy of further biomedical research.

Intestinal or bowel hyperpermeability, so called leaky gut, has been linked to food allergies and some food intolerances. Research is currently focussing on specific conditions and effects of certain food constituents. At present there are a number of ways to limit the increased permeability, but additional studies are required to assess if this approach reduces the prevalence and severity of specific conditions.

References

- Saravacos, George D.; Maroulis, Zacharias B. (2011).Food Process Engineering Operations. CRC Press. pp. 198-199. ISBN 1439877858
- Johnson, Paul (2007). "Lobster". Fish Forever: The Definitive Guide to Understanding, Selecting, and Preparing Healthy, Delicious, and Environmentally Sustainable Seafood. John Wiley & Sons. pp. 163–175. ISBN 978-0-7645-8779-5.
- Havkin-Frenkel, Daphna; Belanger, Faith C. (2011). Handbook of Vanilla Science and Technology. Chichester, UK: Wiley-Blackwell. pp. 142–145. ISBN 978-1-4051-9325-2.
- Sun, Da-Wen (2007). Computer Vision Technology for Food Quality Evaluation. Academic Press. p. 600. ISBN 978-0-12-373642-0.
- Gibbons T, Fuchs GJ (2007). "Chronic enteropathy: clinical aspects". Nestle Nutr Workshop Ser Pediatr Program. Series Set, 2007. 59: 89–101; discussion 102–4. doi:10.1159/000098529. ISBN 3-8055-8194-7. PMID 17245093.
- Zhou, Shi-Sheng (2014). "Excess vitamin intake: An unrecognized risk factor for obesity". World J Diabetes. 5 (1): 1–13. doi:10.4239/wjd.v5.i1.1. ISSN 1948-9358. PMID 24567797.
- Gibson PR; Shepherd SJ. (Feb 2010). "Evidence-based dietary management of functional gastrointestinal symptoms: The FODMAP approach". J Gastroenterol Hepatol. 25 (2): 252–8. doi:10.1111/j.1440-1746.2009.06149.x. PMID 20136989.
- Editorial Staff. "What's the difference between an allergy, an intolerance and a sensitivity?". Healthy Futures. Retrieved 24 August 2010.

Permissions

All chapters in this book are published with permission under the Creative Commons Attribution Share Alike License or equivalent. Every chapter published in this book has been scrutinized by our experts. Their significance has been extensively debated. The topics covered herein carry significant information for a comprehensive understanding. They may even be implemented as practical applications or may be referred to as a beginning point for further studies.

We would like to thank the editorial team for lending their expertise to make the book truly unique. They have played a crucial role in the development of this book. Without their invaluable contributions this book wouldn't have been possible. They have made vital efforts to compile up to date information on the varied aspects of this subject to make this book a valuable addition to the collection of many professionals and students.

This book was conceptualized with the vision of imparting up-to-date and integrated information in this field. To ensure the same, a matchless editorial board was set up. Every individual on the board went through rigorous rounds of assessment to prove their worth. After which they invested a large part of their time researching and compiling the most relevant data for our readers.

The editorial board has been involved in producing this book since its inception. They have spent rigorous hours researching and exploring the diverse topics which have resulted in the successful publishing of this book. They have passed on their knowledge of decades through this book. To expedite this challenging task, the publisher supported the team at every step. A small team of assistant editors was also appointed to further simplify the editing procedure and attain best results for the readers.

Apart from the editorial board, the designing team has also invested a significant amount of their time in understanding the subject and creating the most relevant covers. They scrutinized every image to scout for the most suitable representation of the subject and create an appropriate cover for the book.

The publishing team has been an ardent support to the editorial, designing and production team. Their endless efforts to recruit the best for this project, has resulted in the accomplishment of this book. They are a veteran in the field of academics and their pool of knowledge is as vast as their experience in printing. Their expertise and guidance has proved useful at every step. Their uncompromising quality standards have made this book an exceptional effort. Their encouragement from time to time has been an inspiration for everyone.

The publisher and the editorial board hope that this book will prove to be a valuable piece of knowledge for students, practitioners and scholars across the globe.

Index